# Optical Sensors, Devices and Systems

# Optical Sensors, Devices and Systems

Edited by **Vladimir Latinovic**

**WILLFORD PRESS**

New York

Published by Willford Press,
118-35 Queens Blvd., Suite 400,
Forest Hills, NY 11375, USA
www.willfordpress.com

**Optical Sensors, Devices and Systems**
Edited by Vladimir Latinovic

International Standard Book Number: 978-1-68285-043-5 (Hardback)

# Contents

# Preface

The main aim of this book is to educate learners and enhance their research focus by presenting diverse topics covering this vast field. This is an advanced book which compiles significant studies by distinguished experts. This book addresses successive solutions to the challenges arising in the area of application, along with it; the book provides scope for future developments.

Optical sensors are widely used in different fields for measuring and collecting information from diverse sources. This book attempts to understand the various applications of optical sensors. Some of the important topics elucidated in the book are optical and photonic sensors, signal processing, evaluations and assessments of different applications, etc. The various studies that are constantly contributing towards advancing technologies and evolution of this field are examined in detail. Students, researches and academicians engaged in the field will find this book full of innovative insights and developments in optical sensor technology.

It was a great honour to edit this book, though there were challenges, as it involved a lot of communication and networking between me and the editorial team. However, the end result was this all-inclusive book covering diverse themes in the field.

Finally, it is important to acknowledge the efforts of the contributors for their excellent chapters, through which a wide variety of issues have been addressed. I would also like to thank my colleagues for their valuable feedback during the making of this book.

**Editor**

# Optical Fiber Nanotips Coated with Molecular Beacons for DNA Detection

**Ambra Giannetti [1],\*, Andrea Barucci [1], Franco Cosi [1], Stefano Pelli [1,2], Sara Tombelli [1], Cosimo Trono [1] and Francesco Baldini [1]**

[1] CNR—Institute of Applied Physics "Nello Carrara", Via Madonna del Piano 10, 50019 Sesto Fiorentino (FI), Italy; E-Mails: a.barucci@ifac.cnr.it (A.B.); f.cosi@ifac.cnr.it (F.C.); s.pelli@ifac.cnr.it (S.P.); s.tombelli@ifac.cnr.it (S.T.); c.trono@ifac.cnr.it (C.T.); f.baldini@ifac.cnr.it (F.B.)

[2] Museo Storico della Fisica e Centro Studi e Ricerche Enrico Fermi, Piazza del Viminale 1, 00184 Rome, Italy

\* Author to whom correspondence should be addressed; E-Mail: a.giannetti@ifac.cnr.it

Academic Editor: Ashutosh Tiwari

**Abstract:** Optical fiber sensors, thanks to their compactness, fast response and real-time measurements, have a large impact in the fields of life science research, drug discovery and medical diagnostics. In recent years, advances in nanotechnology have resulted in the development of nanotools, capable of entering the single cell, resulting in new nanobiosensors useful for the detection of biomolecules inside living cells. In this paper, we provide an application of a nanotip coupled with molecular beacons (MBs) for the detection of DNA. The MBs were characterized by hybridization studies with a complementary target to prove their functionality both free in solution and immobilized onto a solid support. The solid support chosen as substrate for the immobilization of the MBs was a 30 nm tapered tip of an optical fiber, fabricated by chemical etching. With this set-up promising results were obtained and a limit of detection (LOD) of 0.57 nM was reached, opening up the possibility of using the proposed nanotip to detect mRNAs inside the cytoplasm of living cells.

**Keywords:** molecular beacon; optical fiber nanotip; nanosensor; optical biosensors; fluorescence; mRNA; DNA; survivin

## 1. Introduction

Usually, information on basic cellular signaling processes associated with diseases is not directly deducible from measurements on an average of a pool of cells. In fact, new information can only be obtained by monitoring the cellular signaling pathways directly inside intact cells [1]. Traditional approaches for studying cells and molecular biology outside living cells (e.g., by using biomolecules purified from cells), while being highly productive, often lose valuable information about cellular mechanisms that can only be observed in their natural environment. Hence, even if challenging, this progress in cellular physiology study requires new detection strategies at the nanoscale level applied to individual cells with large temporal and spatial resolution. In this framework, nanotechnology is becoming a key tool, opening new ways to explore and understand cellular and subcellular mechanisms [2]. In recent years, the advances in nanotechnology have led to the development of nanotools, capable of entering the single cell and detecting target biomolecules inside the living cells if coupled to the specific receptors. Examples of these nanotools are: fluorescent quantum dots, used to visualize dynamic processes in living cells [3–5]; carbon nanotubes or polymeric nanoparticles, used to penetrate the cell membranes as carrier of drugs or for imaging purpose [6–8]; gold nanorods, used for photothermal therapy [9]; and nanoneedles and nanotips which can penetrate the cell membranes with minimal cell damage [10,11].

Among these nanotools, nanoprobes based on fiber-optical nanotips were conceived as integrated nanodevices consisting of a recognition molecule coupled to the optical transducing element (the optical fiber) interfaced to a photo- or spectrometric detection system, aimed to perform diagnosis within single living cells [12–15].

Chemical nanosensors were first developed for monitoring calcium and nitric oxide, and other physiochemical parameters in single cells [16], showing excellent detection limits, as well as photostability, reversibility, and millisecond response. McCulloch *et al.* [17] prepared nanometric optical fiber sensors for intracellular pH measurements, while other groups [18] developed an optical fiber sensor for the measurement of dissolved oxygen based on sol-gel immobilization technology. Vo-Dinh and coworkers have developed nanobiosensors to detect biochemical targets inside living single cells [19–22], while Zhang [23] reported about an integrated device for optical stimulation and spatiotemporal electrical recording of neural activity in light-sensitized brain tissue.

In the field of DNA/RNA analysis in cell, MBs were proposed by Tyagi and Kramer [24–29]. MBs are single-stranded DNA molecules that possess a stem-loop structure commonly named hairpin structure. The loop portion of the molecule can form a double-stranded DNA in the presence of a complementary sequence. MB can be labeled with a fluorophore and a quencher at the two side-ends of the stem, constituted by a more or less long chain of complementary bases, which keeps these two moieties in close proximity to each other. Since the fluorophore is characterized by an emission band, which overlaps the absorption band of the quencher, this proximity causes the fluorescence of the fluorophore to be quenched by energy transfer. The fluorophore fluorescence is restored upon the opening of the stem due to the hybridization of the MB with the target sequence.

A great many fiber nanotips are characterized by a metal coating which allows them to illuminate exactly only the fiber nanotip. Actually, it is very difficult to quantify the exact fraction of light that will reach the nanotip since, below a certain inner diameter, even the lowest guided mode runs into cutoff, where the wave vector becomes imaginary and thus the mode field decays exponentially [30]. However, although the transmitted light fraction can be extremely small at the apex of the nanotip, when applying the nanoprobe for intracellular analysis, the dimensions of a single cell which are in the range 1 μm–10 μm, lead to the fact that the examined part of the nanoprobe will be not only the apex of 30 nm and consequently a larger portion of hundreds of nanometers will be exposed to the intracellular content. The absence of the metal coating on the fiber nanotip implies the excitation of the MB immobilized along the whole fiber nanotip. On the other hand, only the portion of the MB inside the cell which interacts with the intracellular target will emit fluorescence. A partial loss of localization associated with this approach can be accepted for some intracellular application where there is not a strong necessity of localization, if this is accompanied with an increase of the fluorescence signal associated to the excitation of a much larger number of molecules of MB.

In this paper we present preliminary results concerning the use of an optical fiber nanotip coupled to a MB for DNA detection (Figure 1). The solid support chosen as substrate for the immobilization of the MB was a tip of an optical fiber tapered at nanoscale size, intended to be used in the future for mRNA detection inside the cytoplasm of living cells. The nanotip was fabricated by chemical etching, starting from 500 micron—diameter of the multimode optical fiber—down to 30 nm at the tip. Then, the fiber tip was silanized, and the MB was attached via a covalent-binding procedure.

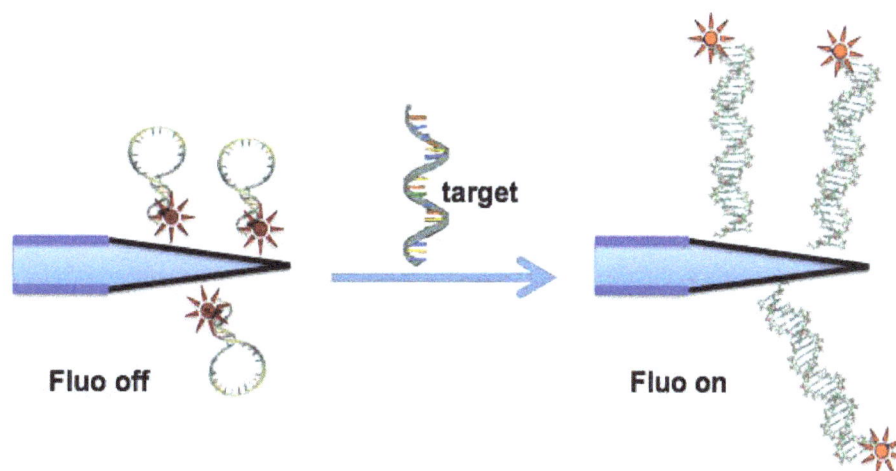

**Figure 1.** Scheme of the sensing mechanism.

In particular, the attention was focused on the mRNA for survivin, a protein highly expressed in most types of cancer. In this case, the MB could act not only as detector of the over-expression of the mRNA for the survivin (diagnosis) but also as the blocking agent of the synthesis of the protein itself (therapy), as we demonstrated with Real-time PCR and western blotting experiments which showed a time-dependent reduction of survivin mRNA and protein after 100 nM-MB treatment, respectively, with the molecular beacon transfected into A375 melanoma cells [31,32].

After the characterization of the sole MB in solution carried out to verify the reliability and the effectiveness of the fluorophore/quencher pair immobilized at the extremities of the single-stranded DNA and at the same time of the good interaction of the MB with its target, a thorough opto-chemical characterization was performed in order to evaluate the possibility of getting a reliable, reproducible and robust system consisting of the MB immobilized onto the fiber nanotip.

## 2. Experimental Section

### 2.1. Chemicals

All the reagents for the preparation of phosphate buffered saline (PBS), 40 mM, pH 7.4; tris hydrochloride (tris-HCl), magnesium chloride ($MgCl_2$) for the preparation of tris-buffer (10 mM tris-HCl, 10 mM $MgCl_2$, pH 8); hydrofluoric acid (HF); hydrochloric acid (HCl); sulfuric acid ($H_2SO_4$); hydrogen peroxide ($H_2O_2$); methanol (MeOH) were purchased from Sigma-Aldrich (Milan, Italy). (3-Aminopropyl)triethoxysilane (APTES) was purchased by abcr GmbH (Karlsruhe, Germany). N-succinimidyl 3-(2-pyridyldithio) propionate (SPDP) was purchased from Pierce (Thermofisher Scientific, Milan, Italy). The molecular beacon for survivin, its complementary target sequence, the non-specific random sequence and the labeled linear probe were purchased from IBA (Gottingen, Germany).

### 2.2. Molecular Beacon and Target Sequences

The sequence for the survivin molecular beacon [31] was chosen among several published sequences [33–36] for its better performances in terms of folding properties and greater reactivity with the target sequence. A different fluorophore-quencher pair, ATTO647N ($\lambda_{abs}$ 644 nm, $\lambda_{em}$ 669 nm) and BlackBerry® Quencher 650 ($\lambda_{max} \sim$ 650 nm, useful absorbance between 550 and 750 nm), was chosen with respect to the one used for the published MBs in order to obtain a greater quantum yield (QY = 65 of the ATTO647N instead of 0.28 of the Cy5). The sequences of the different oligonucleotides (MB, specific target, random sequence and linear probe) are reported below:

MB                    5'-(ATTO647N)CGACGGAGAAAGGGCTGCCACGXCG(BBQ)-3' X=C6-dT Thio
Target                5'-CCCCTGCCTGGCAGCCCTTTCTCAAGGACC-3'
Random sequence 5'-ATCGGTGCGCTTGTCG-3'
Linear probe       5'-(ATTO647N)GAGAAAGGGCTGCCA(Thiol)-3'

All the measurement carried out with both the MB in solution and the MB immobilized on the fiber tip were carried out at room temperature.

### 2.3. Optical Fiber Nanotip Preparation

The optical fiber nanotip was manufactured by a dynamic chemical etching method [37,38], by mechanically rotating and dipping a 3 M silica optical fiber, (core diameter 480 micron) in a chemical etching solution (aqueous hydrofluoric acid) covered with a protection layer.

Using different dynamic regimes of the mechanical movements during the chemical etching process, it was possible to vary the cone angle, the shape and the roughness of the nanoprobes. It was found that the tip profiles were determined by the nonlinear dynamic evolution of the meniscus of the etching solution near the fiber. Different regimes can be generated by changing the ratio between the angular velocities and the ratio between the radii of optical fiber and vial, ranging from laminar flow to the onset of chaotic flow. The type of flow of the viscous HF solution caused by the rotation of the vial and the optical fiber can be described in the framework of the Taylor-Couette flow (TCF) theory [39,40].

Moreover, by an accurate control of the extraction speed of the fiber from the HF solution, the length and the angle of the nanotip were precisely controlled, and the capillarity effects at the nanoscale were minimized by limiting strong friction on the optical fiber nanotip in the final part of the etching process.

SEM pictures and AFM measurements have been performed on different nanotips showing our ability to reproduce typical tip features, such as short taper length (~200 μm), large cone angle (from 15° to 40°), small probe tip dimension (less than 30 nm), and roughness below 10 nm. The geometrical characterization demonstrated that, with this method, a high yield of reproducible nanotips can be obtained, overcoming some drawbacks of conventional etching techniques. The nanometric roughness is the key point to keep the scattering, which can perturb the measurements, at the lowest levels. Figure 2 shows an example of nanotip obtained using the fabrication method described above.

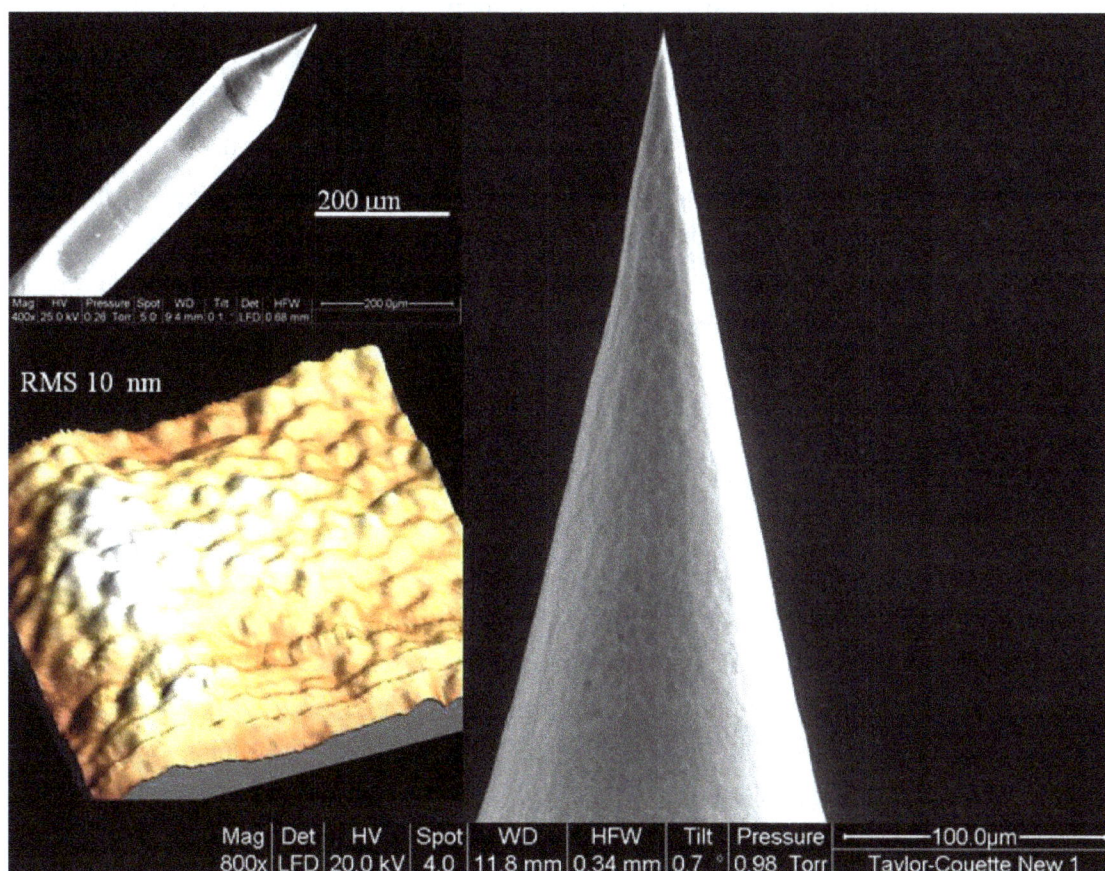

**Figure 2.** Example of a nanotip obtained using the fabrication method described in Section 2.3. In the left top a general view showing a taper length of about 200 μm and a large cone angle (about 40°), while in the left bottom an AFM image (10 μm × 10 μm) is reported, with an estimated roughness of about 10 nm. On the right panel is reported a zoom on the distal end, showing an apex angle of about 24°.

In particular, in the left top part a general view of the short taper length and the large cone angle (about 40° in this example) of the nanotip is shown, while in the left bottom an AFM image (10 μm × 10 μm) is reported, with an estimated roughness of the tip of about 10 nm. On the right panel is reported a zoom on the distal end, showing the apex angle (about 24°).

The glass nanotip was dipped in piranha solution ($H_2SO_4$:$H_2O_2$ 7:3) for 30 min to carry out a strong cleaning and to activate the silanol groups present on the surface. Then the tapered fiber nanotip was functionalized by a 2-hour long silanization procedure using 5% APTES in MeOH/$H_2O$ (1:1) in order to achieve the amino groups onto the surface of the nanotip. After several rinsing cycles in MeOH and deionized water, the nanotip was cured in oven at 100 °C for 1 h.

SPDP was used as crosslinker between the -$NH_2$ groups available onto the nanotip and the -SH groups of the linear probe or of the MB. After the treatment with a solution of SPDP 4 mM in PBS at pH 7.4 for 30 min, the nanotip was rinsed in PBS and then in tris-buffer. As already reported by Tombelli *et al* [41], the cross-linker SPDP could be used to form a cleavable disulphide bond to detach the MB from the delivery vehicle inside the cytoplasm. This could be an advantage if the aim is to release the MB inside the cells, and, consequently the nanotip is used simply as the delivery tool of the MB as therapeutic molecule inside the cell. On the other hand, if the probe is used as imaging tool, other cross-linkers, such as succinimidyl-4-(N-maleimidomethyl)cyclohexane-1-carboxylate (SMCC) should be considered in order to avoid any loss of signal caused by the MB detachment from the fiber nanotip.

Before its immobilization, the MB was heated for 5 min in water bath at about 70–80 °C (slightly higher than the melting temperature) and then left to slowly equilibrate down to room temperature in order to optimize the hairpin formation. The immobilization of 1 μM MB (in tris-buffer), or of 1 μM linear probe, onto the nanotip was performed over night at 4 °C.

### 2.4. The Optical Measurements

Scanning Electron Microscope (SEM, Quanta-200 FEI, Hillsboro, OR, USA) was used to characterize the geometrical features of the nanotips, whereas the surface roughness was checked by Atomic Force Microscopy (AFM, prototype developed by Pini *et al.* [42]).

An in-house optical set-up, illustrated in Figure 3, was used to characterize the fluorescent nanoprobes. Fluorescence measurements on fiber nanotips were carried out in a cuvette using, for the MB excitation, a LDH-P-C-635B laser diode (PicoQuant, Berlin, Germany) emitting at 635 nm filtered with a bandpass interference filter (FL635-10, ThorLabs, Newton, NJ, USA). The emitted fluorescence was collected by means of a GRIN lens coupled with a multimode optical fiber (diameter 200 μm), aligned with the fiber nanotip and then guided to an optical high-pass filter Thorlabs FEL0650, in order to filter the excitation light scattered out from the fiber tip, and finally acquired by a Shamrock 303i spectrograph (Andor, Belfast, United Kingdom). The cuvette was fixed inside a suitable holder onto a manual labjack (model 271 Labjack, Newport, Irvine, CA, USA) and the fiber was immersed into the cuvette from the top. In this way, it was possible, after the optical alignment, to move down the cuvette, to empty and fill it or to exchange the cuvette, and to move up again it to the exact same position, without changing the alignment of the nanotip and GRIN lens.

The experimental set-up for fluorescence measurement of the molecular beacon in solution was very similar to the one described in Figure 3, with the difference that the fluorescent solution into the cuvette was excited directly with the collimated laser beam, without the objective, and collected with the GRIN lens in normal direction with respect to the excitation laser beam direction. The measurements in solution were performed without the optical high-pass filter Thorlabs FEL0650, because of the better signal (fluorescence) to noise (scattered excitation light) ratio of this measurement configuration.

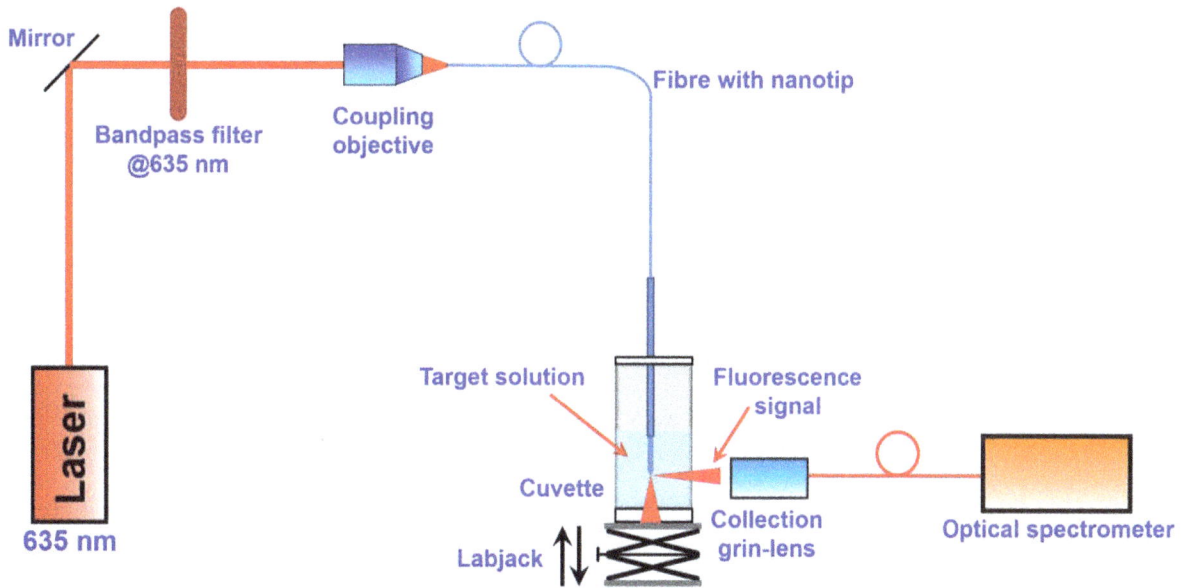

**Figure 3.** Optical experimental set-up.

## 3. Results and Discussion

### 3.1. MB Characterization in Solution

The functionality of the MB was firstly investigated in buffer solution by hybridization with the specific target at different incubation times. Different buffers and different culture media were already tested and reported, such as Dulbecco's modified Eagle's Medium (DMEM) [43,44]. Even if tris-buffer is not mimicking the environmental condition within the cell, it is the one which provides the best performances of the molecular beacon in solution.

The MB was characterized by recording the fluorescence ($\lambda_{ex}$ 635 nm) of the MB 100 nM in Tris buffer alone and after 1, 3 and 5 h of incubation with the specific target 0.2 μM (Figure 4). A final 10-times increase of the MB fluorescence in presence of the target was recorded with the signal increasing with time from 1 to 5 h. The characterization on such a long time, much longer than the interaction time between the MB and the target, was carried out in order to verify the stability of the molecular beacon and, consequently, of its emitted fluorescence also after that the interaction took place.

### 3.2. MB and Linear Probe Characterization onto the Nanotip

Before working with the molecular beacon, the fluorescent linear probe was immobilized onto the nanotip in order to evaluate the actual possibility of sensing the presence of a labeled oligonucleotide by direct excitation of the oligonucleotide fluorophore by means of the same fiber nanotip. After the probe immobilization, the fluorescence of the ATTO647N fluorophore labeling the probe, was measured ($\lambda_{ex}$ 635 nm). Figure 5 shows the fluorescence spectrum of the linear probe immobilized onto the nanotip. The fluorescence peak shown in the Figure is split into two peaks, at 660 nm and 680 nm due to the not uniform transmission spectrum of the Thorlabs FEL0650 high-pass filter.

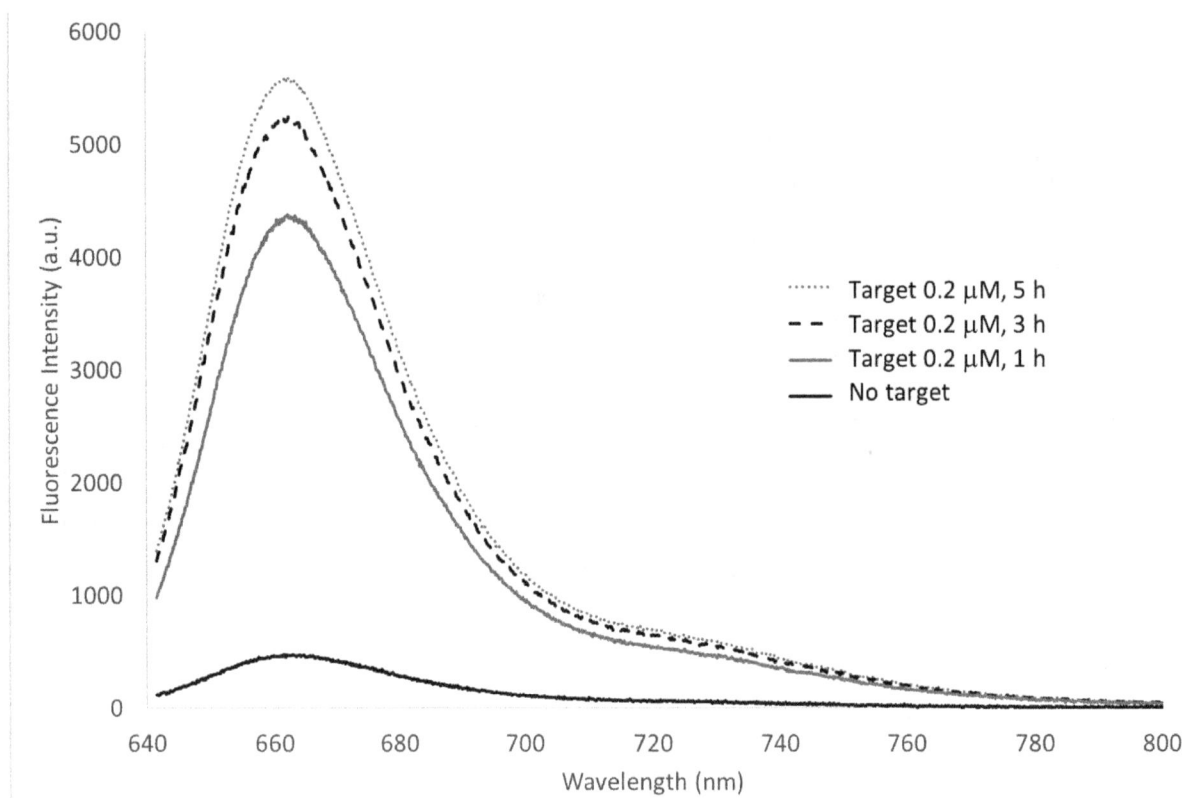

**Figure 4.** Fluorescence spectra of the molecular beacon (0.1 μM) after incubation for 1, 3 and 5 h with the specific target 0.2 μM. $\lambda_{ex}$ 635 nm, exposure time 1 s.

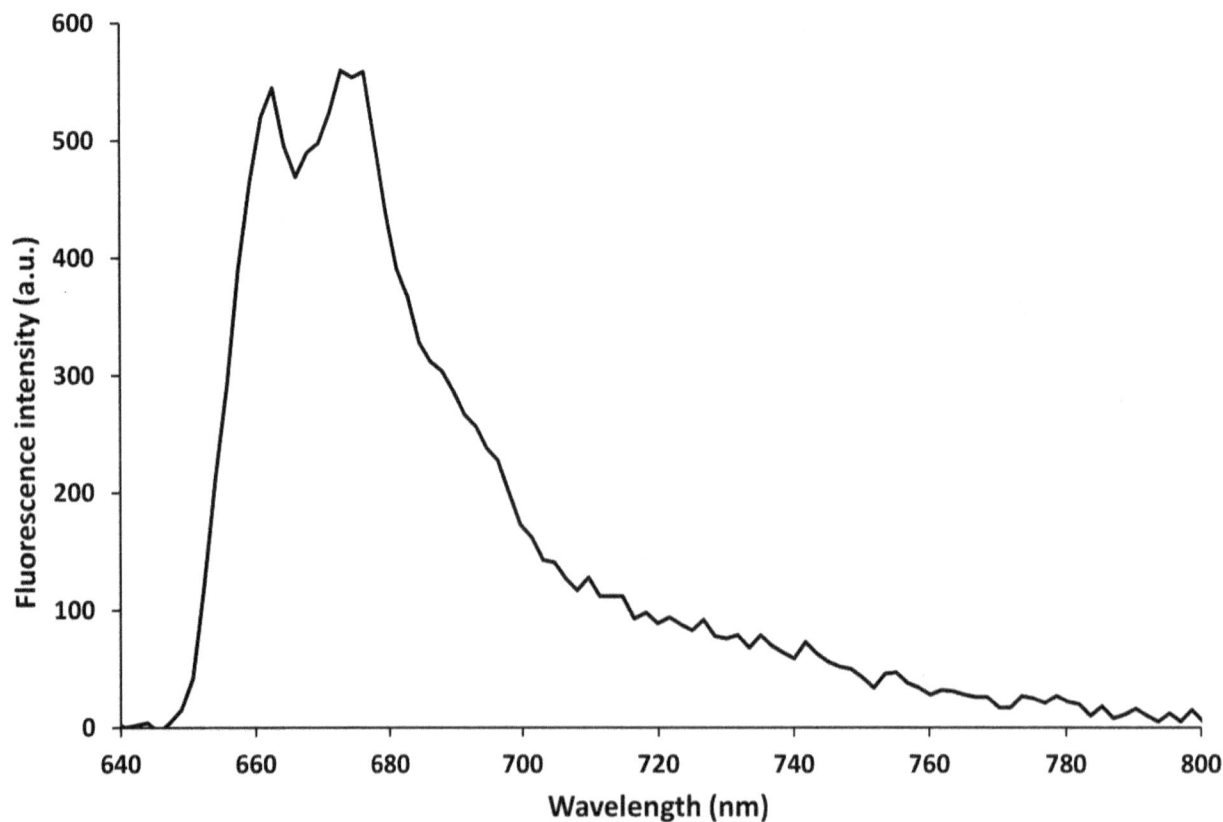

**Figure 5.** Fluorescence spectra of the linear probe immobilized onto the nanotip. $\lambda_{ex}$ 635 nm, exposure time 1 s.

The fluorescence signal of the probe was monitored for 30 min in order to evaluate the fluorescence stability: the average fluorescence intensity in 30 min was of 11,300 (a.u.) with a standard deviation of 436 (a.u.) leading to a signal variability of 3.8%.

In Figure 6 the fluorescence spectrum of the multimode fiber functionalized with the molecolar beacon is shown. In particular, the nanotip was immersed in pure buffer (Tris), in a solution of the random sequence and in two solutions of the specific target 1 and 10 µM and the fluorescence spectra were recorded after 30 min.

The performance of the fiber nanotip to act as sensor was evaluated by monitoring the fluorescence signal as a function of time, when exposed to different concentrations of the target. Its specificity when interacting with the random sequence was also investigated (Figure 7). In particular, after an initial stabilization of the MB fluorescence in buffer, the coated nanotip was first immersed in a solution 1 µM of the random sequence and the fluorescence was recorded for 30 min. Thereafter, the same nanotip was consecutively incubated for the same time in target solutions at different concentrations. As can be seen from the figure, only a 3.5% fluorescence increase was observed in presence of the random sequence, whereas the increase with only 0.01 µM of the target was of about 13%. Each point of the graph is evaluated as the sum of the optical intensity detected by the optical spectrograph between 650 and 720 nm.

The curve in Figure 7 well testifies for the feasibility of the MB-coated nanotip to act as a sensor, providing also information on the kinetics of the interaction between the MB and the target. The nanotip was exposed to the different solutions (random and target solutions) for 30 min in order to perform an opto-chemical characterization, even if the timing needed was showed to be minor than 10 min.

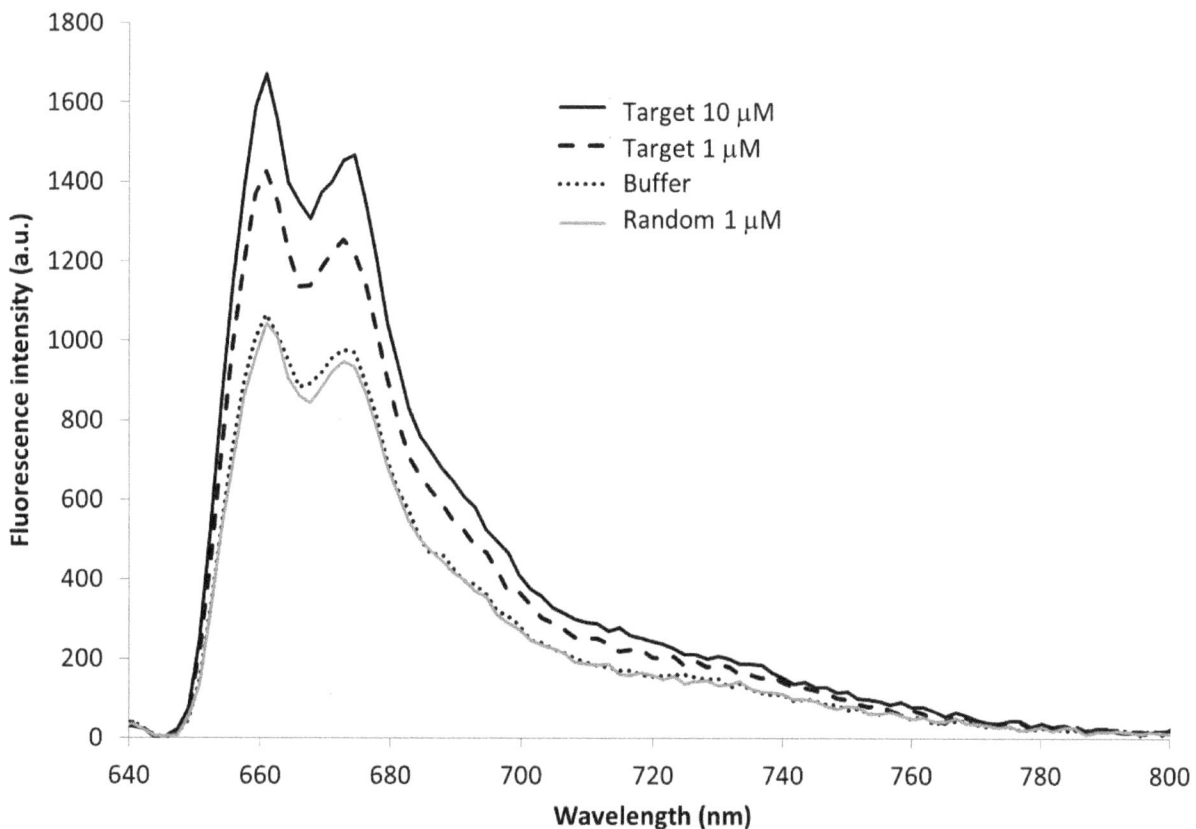

**Figure 6.** MB immobilized onto the nanotip, characterization in presence of the target (1 µM and 10 µM) and of the random sequence (1 µM). $\lambda_{ex}$ 635 nm, exposure time 1 s.

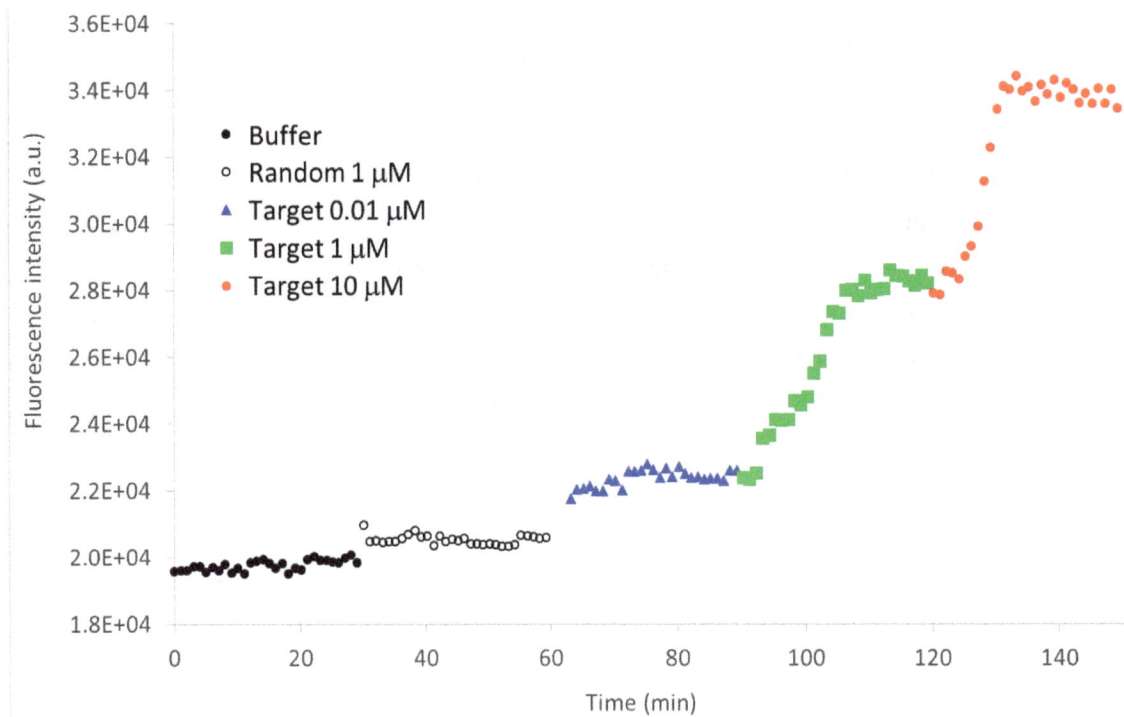

**Figure 7.** Fluorescence signal (*vs.* time) during the interaction of the MB immobilized onto the nanotip with the random sequence, and the target at different concentrations (0.01 µM, 1 µM and 10 µM).

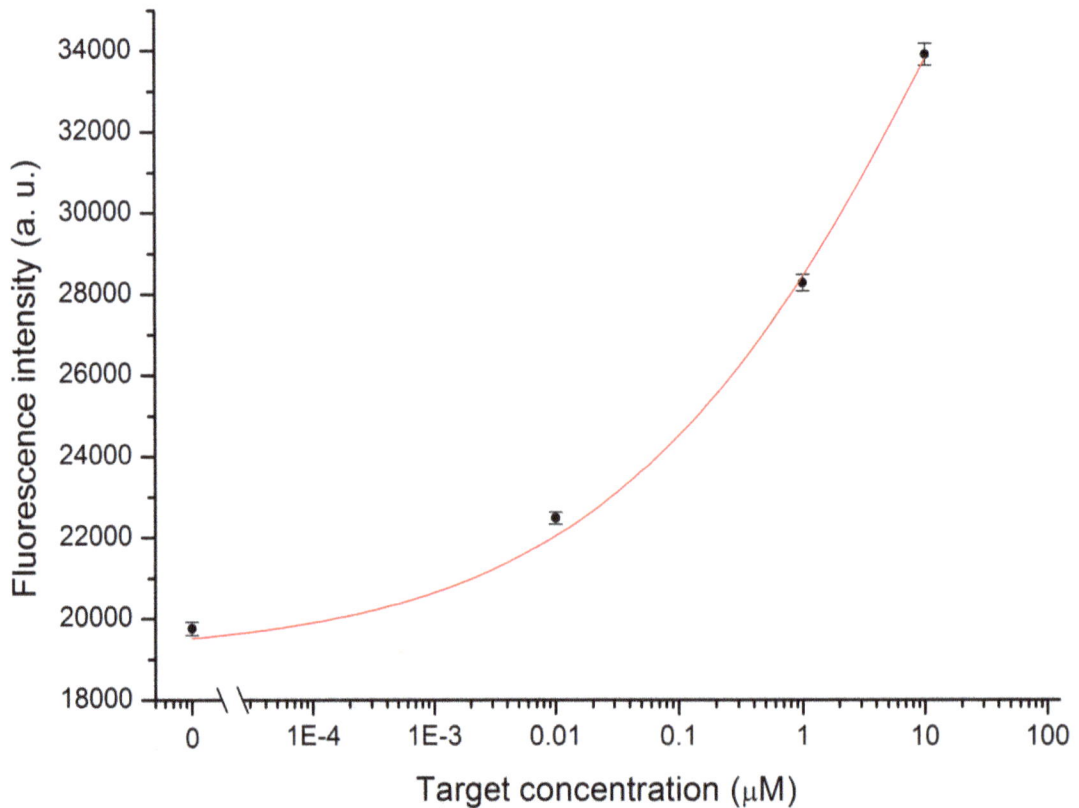

**Figure 8.** Calibration curve obtained from the interaction of the MB-coated nanotip with increasing concentrations of target.

Figure 8 shows the calibration curve obtained by averaging the experimental data at the plateau of the binding curve and calculating the related standard deviation for each target concentration. A Limit of Detection (LOD) of 0.57 nM for the proposed sensor was calculated from the calibration curve considering three times the greater difference between the experimental data and the fitting curve (maximum of residuals).

The limit of detection achieved opens up the possibility of using the proposed nanotip to detect mRNAs inside the cytoplasm of living cells. In fact, the concentration of specific mRNAs has been estimated to be in the nM range [10], although it is quite difficult to estimate the mRNA concentration inside the cytoplasm since it is not homogeneously distributed, but it has preferential locations. In the case of survivin mRNA, which is the molecule of interest, preferential localization is at the perinuclear position [34].

By investigating the possibility of regenerating the nanotip for multiple use by the use of HCl, we discovered that it is possible to increase the performance of the sensor. Figure 9 shows the comparison between the response of the fiber nanotip to the 30 min exposure to 1 μM target solution (grey tracing) shown in Figure 7 (green tracing) and the response to the same target concentration after having exposed the fiber nanotip to HCl 2 mM for 1 min (black tracing). In this graph, the baselines of the two curves were both moved to the value of zero for a better comparison. After the HCl treatment, the nanotip was thoroughly washed in tris-buffer and left to equilibrate in this buffer until a stable fluorescence signal was reached, before exposing it to the target solution.

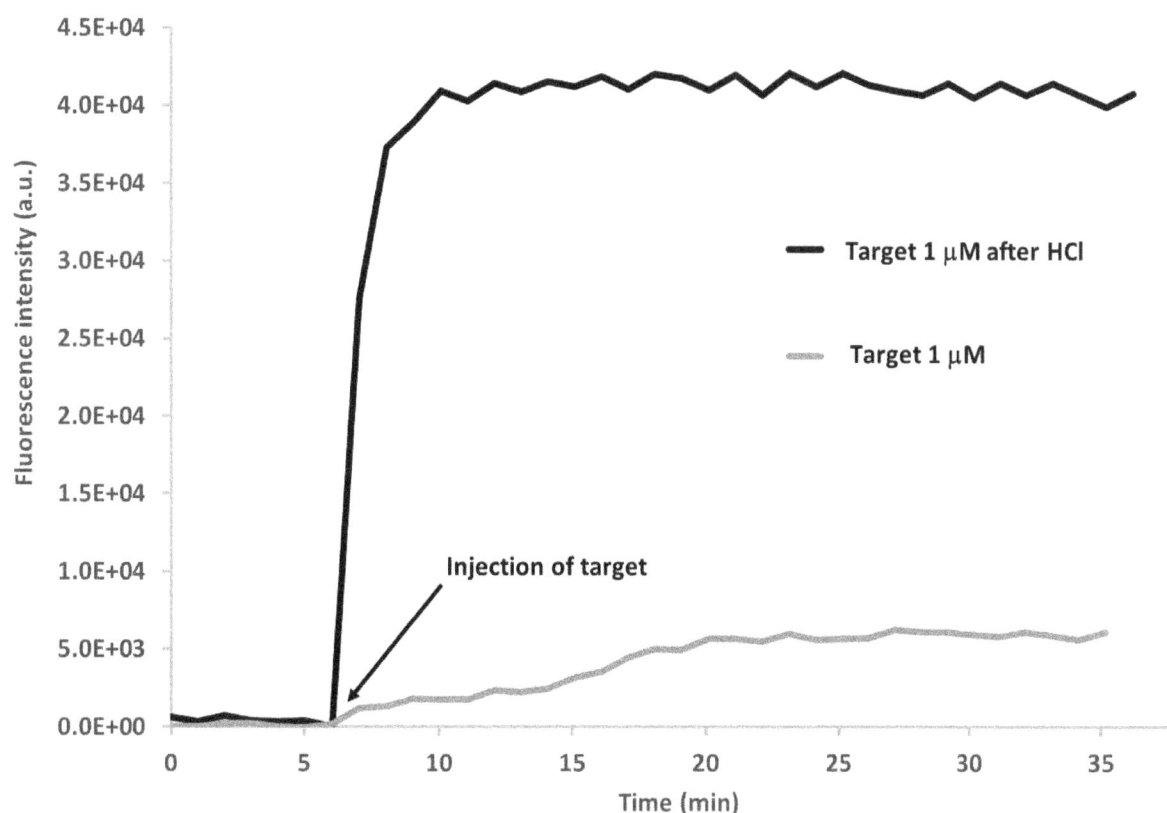

**Figure 9.** Kinetics of the fluorescence signal of the MB onto the nanotip after incubation with the target 1 μM without any previous treatment and after the treatment with HCl 2 mM.

An 8-fold increase in the fluorescence signal due to the interaction of the MB with the target can be observed. This behaviour can be ascribed to the unfolding effect of HCl which is able to open the hairpin structure of the MB and the subsequent refolding of the beacon hairpin structure in a more efficient manner with respect to the untreated beacon which could be also folded around the nanotip and consequently hindered towards the hybridisation with the target. This fact can provide the explanation of the faster kinetics which is observed after the HCl treatment, with the plateau reached after only 4 min, instead of 15 min. The use of HCl 1–2 mM, among others regenerating solutions, is quite common in DNA biosensor regeneration [45–47]. The demonstration that the MB is not degraded by this treatment is given by the high and fast fluorescent increase from PBS (target zero) to the target 1 µM, as shown in Figure 8. The HCl treatment is therefore essential in order to achieve a correct and ready-to-interact MB layer on the fiber nanotip immediately after its immobilization which can occur also with secondary structures hampered towards the proper interaction with the target. The same treatment could also be used in the regeneration of the sensor after the interaction with the target in order to remove the target from the MB and make it fold again in its closed structure ready for a new hybridisation cycle.

## 4. Conclusions

The present paper describes the manufacture and optical characterization of an optical fiber nanotip coated with molecular beacons potentially capable to perform intracellular measurements. The optical characterization showed the good performance of the obtained nanotip in terms of selectivity and sensitivity, and a limit of detection of 0.57 nM was achieved. The treatment with HCl, besides allowing a regeneration of the nanotip, is able to provide a better performance of the molecular beacon with a further decrease of the limit of detection.

## Acknowledgments

The authors wish to thank Bruno Tiribilli for the acquisition of the AFM images, Giancarlo C. Righini for the fruitful discussion during the whole course of this study and the project FIRB ITALNANONET (RBPR05JH2P) for financial support.

## Author Contributions

Ambra Giannetti proposed the work, designed and supervised the experiments; Ambra Giannetti and Sara Tombelli performed the chemical/biochemical experiments; Andrea Barucci and Franco Cosi performed the fabrication of the nanotip; Cosimo Trono constructed the optical set-up and conduced the optical measurements; all authors wrote the paper and provided comments on it.

## Conflicts of Interest

The authors declare no conflict of interest.

# References

1. Vo-Dinh, T. Nanosensing at the single cell level. *Spectrochim. Acta Part B At. Spectrosc.* **2008**, *63*, 95–103.

2. Zheng, X.T.; Hu, W.; Wang, H.; Yang, H.; Zhou, W.; Li, C.M. Bifunctional electro-optical nanoprobe to real-time detect local biochemical processes in single cells. *Biosens. Bioelectron.* **2011**, *26*, 4484–4490.

3. Dahan, M.; Lévi, S.; Luccardini, C.; Rostaing, P.; Riveau, B.; Triller, A. Diffusion dynamics of glycine receptors revealed by single-quantum dot tracking. *Science* **2003**, *302*, 442–445.

4. Pathak, S.; Cao, E.; Davidson, M.C.; Jin, S.; Silva, G.A. Quantum Dot Applications to Neuroscience: New Tools for Probing Neurons and Glia. *J. Neurosci.* **2006**, *26*, 1893–1895.

5. Ma, Q.; Lin, Z.; Yang, N.; Li, Y.; Su, X.-G. A novel carboxymethyl chitosan–quantum dot-based intracellular probe for $Zn^{2+}$ ion sensing in prostate cancer cells. *Acta Biomater.* **2014**, *10*, 868–874.

6. Duchi, S.; Sotgiu, G.; Lucarelli, E.; Ballestri, M.; Dozza, B.; Santi, S.; Guerrini, A.; Dambruoso, P.; Giannini, S.; Donati, D.; *et al.* Mesenchymal stem cells as delivery vehicle of porphyrin loaded nanoparticles: Effective photoinduced *in vitro* killing of osteosarcoma. *J. Control Release* **2013**, *168*, 225–237.

7. Giannetti, A.; Tombelli, S.; Baldini, F. Oligonucleotide optical switches for intracellular sensing. *Anal. Bioanal. Chem.* **2013**, *405*, 6181–6196.

8. Sokolova, V.; Epple, M. Inorganic Nanoparticles as Carriers of Nucleic Acids into Cells. *Angew. Chem. Int. Ed.* **2008**, *47*, 1382–1395.

9. Ratto, F.; Matteini, P.; Centi, S.; Rossi, F.; Pini, R. Gold nanorods as new nanochromophores for photothermal therapies. *J. Biophoton.* **2010**, *4*, 64–73.

10. Kihara, T.; Yoshida, N.; Kitagawa, T.; Nakamura, C.; Nakamura, N.; Miyakea, J. Development of a novel method to detect intrinsic mRNA in a living cell by using a molecular beacon-immobilized nanoneedle. *Biosens. Bioelectron.* **2010**, *26*, 1449–1454.

11. Yum, K.; Wang, N.; Yu, M.-F. Nanoneedle: A multifunctional tool for biological studies in living cells. *Nanoscale* **2010**, *2*, 363–372.

12. Dufrêne, Y.; Garcia-Parajo, M.F. Recent progress in cell surface nanoscopy: Light and force in the near-field. *Nano Today* **2012**, *7*, 390–403.

13. Zheng, X.T.; Li, C.M. Single living cell detection of telomerase over-expression for cancer detection by an optical fibre nanobiosensor. *Biosen. Bioelectron.* **2010**, *25*, 1548–1552.

14. Zheng, X.T.; Yang, H.B.; Li, C.M. Optical Detection of Single Cell Lactate Release for Cancer Metabolic Analysis. *Anal. Chem.* **2010**, *82*, 5082–5087.

15. Vitol, E.A.; Orynbayeva, Z.; Friedman, G.; Gogotsi, Y. Nanoprobes for intracellular and single cell surface-enhanced Raman spectroscopy (SERS). *J. Raman Spectrosc.* **2012**, *43*, 817–827.

16. Brasuel, M.; Kopelman, R.; Kasman, I.; Miller, T.J.; Philbert, M.A. Ion Concentrations in Live Cells from Highly Selective Ion Correlations Fluorescent Nano-Sensors for Sodium. *IEEE Proc.* **2002**, *1*, 288–292.

17. McCulloch, S.; Uttamchandani, D. Development of a fibre optic micro-optrode for intracellular pH measurements. *IEE Proc. Optoelectron.* **1997**, *144*, 162–167.

18. Xu, H.; Aylott, J.W.; Kopelman, R.; Miller, T.J.; Philbert, M.A. A Real-Time Ratiometric Method for the Determination of Molecular Oxygen Inside Living Cells Using Sol−Gel-Based Spherical Optical Nanosensors with Applications to Rat C6 Glioma. *Anal. Chem.* **2001**, *73*, 4124–4133.

19. Vo-Dinh, T.; Alarie, J.P.; Cullum, B.M.; Griffin, G.D. Antibody-based nanoprobe for measurement of a fluorescent analyte in a single cell. *Nat. Biotechnol.* **2000**, *18*, 764–767.

20. Vo-Dinh, T.; Kasili, P. Fibre-optic nanosensors for single-cell monitoring. *Anal. Bioanal. Chem.* **2005**, *382*, 918–925.

21. Vo-Dinh, T.; Kasili, P.; Wabuyele, M. Nanoprobes and nanobiosensors for monitoring and imaging individual living cells. *Nanomed. Nanotechnol. Biol. Med.* **2006**, *2*, 22–30.

22. Zhang, Y.; Dhawan, A.; Vo-Dinh, T. Design and Fabrication of Fibre-Optic Nanoprobes for Optical Sensing. *Nanoscale Res. Lett.* **2011**, *6*, doi:10.1007/s11671-010-9744-5.

23. Zhang, J.; Laiwalla, F.; Kim, J.A.; Urabe, H.; Van Wagenen, R.; Song, Y.-K.; Connors, B.W.; Zhang, F.; Deisseroth, K.; Nurmikko, A.V. Integrated device for optical stimulation and spatiotemporal electrical recording of neural activity in light-sensitized brain tissue. *J. Neural Eng.* **2009**, *6*, doi:10.1088/1741-2560/6/5/055007.

24. Tyagi, S.; Kramer, F.R. Molecular beacons: Probes that fluoresce upon hybridization. *Nat. Biotechnol.* **1996**, *14*, 303–308.

25. Liu, X.; Farmerie, W.; Schuster, S.; Tan, W. Molecular Beacons for DNA Biosensors with Micrometer to Submicrometer Dimensions. *Anal. Biochem.* **2000**, *283*, 56–63.

26. Monroy-Contreras, R.; Vaca, L. Molecular Beacons: Powerful Tools for Imaging RNA in Living Cells. *J. Nucleic Acids* **2011**, *2011*, doi:10.4061/2011/741723.

27. Santangelo, P.J. Molecular beacons and related probes for intracellular RNA imaging. *Wiley Interdiscip. Rev. Nanomed. Nanobiotechnol.* **2010**, *2*, 11–19.

28. Wang, Q.; Chen, L.; Long, Y.; Tian, H.; Wu, J. Molecular Beacons of Xeno-Nucleic Acid for Detecting Nucleic Acid. *Theranostics* **2013**, *3*, 395–408.

29. Boutorine, A.S.; Novopashina, D.S.; Krasheninina, O.A.; Nozeret, K.; Venyaminova, A.H. Fluorescent Probes for Nucleic Acid Visualization in Fixed and Live Cells. *Molecules* **2013**, *18*, 15357–15397.

30. Hecht, B.; Sick, B.; Wild, U.P.; Deckert, V.; Zenobi, R. Scanning near-field optical microscopy with aperture probes: Fundamentals and applications. *J. Chem. Phys.* **2000**, *112*, 7761–7774.

31. Carpi, C.; Fogli, S.; Giannetti, A.; Adinolfi, B.; Tombelli, S.; Da Pozzo, E.; Vanni, A.; Martinotti, E.; Martini, C.; Breschi, M.C.; *et al.* Theranostic properties of a survivin-directed molecular beacon in human melanoma cells. *PLoS ONE* **2014**, *11*, doi:10.1371/journal.pone.0114588.

32. Adinolfi, B.; Carpi, S.; Giannetti, A.; Nieri, P.; Pellegrino, M.; Sotgiu, G.; Tombelli, S.; Trono, C.; Varchi, G.; Baldini, F. Complex nanostructures based on oligonucleotide optical switches and nanoparticles for intracellular mRNA sensing and silencing. *Proc. Eng.* **2014**, *87*, 751–754.

33. Nitin, N.; Santangelo, P.J.; Kim, G.; Nie, S.; Bao, G. Peptide-linked molecular beacons for efficient delivery and rapid mRNA detection in living cells. *Nucleic Acids Res.* **2004**, *32*, doi:10.1093/nar/gnh063.

34. Santangelo, P.J.; Nix, B.; Tsourkas, A.; Bao, G. Dual FRET molecular beacons for mRNA detection in living cells. *Nucleic Acids Res.* **2004**, *32*, doi:10.1093/nar/gnh062.

35. Qiao, G.; Gao, Y.; Li, N.; Yu, Z.; Zhuo, L.; Tang, B. Simultaneous Detection of Intracellular Tumor mRNA with Bi-Color Imaging Based on a Gold Nanoparticle/Molecular Beacon. *Chem. Eur. J.* **2011**, *17*, 11210–11215.

36. Xue, Y.; An, R.; Zhang, D.; Zhao, J.; Wang, X.; Yang, L.; He, D. Detection of survivin expression in cervical cancer cells using molecular beacon imaging: New strategy for the diagnosis of cervical cancer. *Eur. J. Obstet. Gynecol. Reprod. Biol.* **2011**, *159*, 204–208.

37. Barucci, A.; Cosi, F.; Pelli, S.; Soria, S.; Nunzi Conti, G.; Giannetti, A.; Righini, G. Method of Fabricating Structures, Starting from Material Rods. Patent Pending PCT/EP2014/071743, 10 October 2014.

38. Barucci, A.; Cosi, F.; Giannetti, A.; Pelli, S.; Griffini, D.; Insinna, M.; Salvadori, S.; Tiribilli, B.; Righini, G.C. Optical fibre nanotips fabricated by a dynamic chemical etching for sensing applications. *J. Appl. Phys.* **2015**, *117*, doi:10.1063/1.4906854.

39. Taylor, G.I. Stability of a Viscous Liquid Contained between Two Rotating Cylinders. *Philosoph. Trans. R. Soc. London Ser. A Contain. Pap. Math. Phys. Charact.* **1923**, *223*, 289–343.

40. Andereck, C.D.; Liu, S.S.; Swinney, H.L. Flow regimes in a circular Couette system with independently rotating cylinders. *J. Fluid Mech.* **1986**, *164*, 155–183.

41. Tombelli, S.; Ballestri, M.; Giambastiani, G.; Giannetti, A.; Guerrini, A.; Sotgiu, G.; Trono, C.; Tuci, G.; Varchi, G.; Baldini, F. Oligonucleotide switches and nanomaterials for intracellular mRNA sensing. *Proc. SPIE* **2013**, *8798*, doi:10.1117/12.2033185.

42. Pini, V.; Tiribilli, B.; Gambi, C.M.C.; Vassalli, M. Dynamical characterization of vibrating AFM cantilevers forced by photothermal excitation. *Phys. Rev. B.* **2010**, *81*, doi:10.1103/PhysRevB.81.054302.

43. Giannetti, A.; Baldini, F.; Ballestri, M.; Ghini, G.; Giambastiani, G.; Guerrini, A.; Sotgiu, G.; Tombelli, S.; Trono, C.; Tuci, G.; *et al.* Intracellular nanosensing and nanodelivery by PMMA nanoparticles. *Lect. Notes Electri. Eng.* **2014**, *162*, 69–75.

44. Giannetti, A.; Tombelli, S.; Trono, C.; Ballestri, M.; Giambastiani, G.; Guerrini, A.; Sotgiu, G.; Tuci, G.; Varchi, G.; Baldini, F. Intracellular delivery of molecular beacons by PMMA nanoparticles and carbon nanotubes for mRNA sensing. *Proc. SPIE.* **2013**, *8596*, doi:10.1117/12.2007391.

45. Lechuga, L.M. New frontiers in optical biosensing. In Proceedings of the 13th European Conference on Integrated Optics (ECIO 2007), Copenhagen, Denmark, 25–27 April 2007.

46. Mannelli, I.; Minunni, M.; Tombelli, S.; Wang, R.; Spiriti, M.M.; Mascini, M. Direct immobilization of DNA probes for the development of affinity biosensors. *Bioelectrochemistry* **2005**, *66*, 129–138.

47. Dos Santos Riccardi, C.; Yamanaka, H.; Josowicz, M.; Kowalik, J.; Mizaikoff, B.; Kranz, C. Label-free DNA detection based on modified conducting polypyrrole films at microelectrodes. *Anal. Chem.* **2006**, *78*, 1139–1145.

# Reversible NO₂ Optical Fiber Chemical Sensor Based on LuPc₂ Using Simultaneous Transmission of UV and Visible Light

**Antonio Bueno [1,*], Driss Lahem [2], Christophe Caucheteur [1] and Marc Debliquy [3]**

[1]  Service d'Electromagnétisme et de Télécommunications, Université de Mons, Boulevard Dolez 31, 7000 Mons, Belgium; E-Mail: christophe.caucheteur@umons.ac.be

[2]  Materia Nova, Materials R&D Centre, Parc Initialis, Avenue Nicolas Copernic 1, 7000 Mons, Belgium; E-Mail: driss.lahem@materianova.be

[3]  Service de Science des Matériaux, Université de Mons, Rue de l'Epargne 56, 7000 Mons, Belgium; E-Mail: marc.debliquy@umons.ac.be

*  Author to whom correspondence should be addressed; E-Mail: antonio.buenomartinez@umons.ac.be

Academic Editor: Ki-Hyun Kim

**Abstract:** In this paper, an NO₂ optical fiber sensor is presented for pollution monitoring in road traffic applications. This sensor exploits the simultaneous transmission of visible light, as a measurement signal, and UV light, for the recovery of the NO₂ sensitive materials. The sensor is based on a multimode fiber tip coated with a thin film of lutetium bisphthalocyanine (LuPc₂). The simultaneous injection of UV light through the fiber is an improvement on the previously developed NO₂ sensors and allows the simplification of the sensor head, rendering the external UV illumination of the film unnecessary. Coatings of different thicknesses were deposited on the optical fiber tips and the best performance was obtained for a 15 nm deposited thickness, with a sensitivity of 5.02 mV/ppm and a resolution of 0.2 ppb in the range 0–5 ppm. The response and recovery times are not dependent on thickness, meaning that NO₂ does not diffuse completely in the films.

**Keywords:** optical fiber sensor; nitrogen dioxide; lutetium bisphthalocyanine; traffic pollution

# 1. Introduction

Nowadays, there is growing concern about toxic gases and air contaminants due to the increasing consumption of fossil fuels. Indeed, the combustion of these fuels leads to a massive release of nitrogen oxides ($NO_x$) and other pollutants, such as $SO_2$, CO and $CO_2$. This can cause environmental issues such as the greenhouse effect, acid rain, production of ozone in low atmosphere and air pollution (known as "smog") [1]. In recent decades, the increase of road traffic has made the emissions of internal combustion engines the main source of $NO_x$, which is one of the main toxic gases that can cause respiratory and coronary diseases [2]. For this reason, the need for gas sensors suitable for monitoring and controlling $NO_2$ has arisen. Numerous efforts have been made in the development of $NO_2$ gas sensors.

Among the different technologies, optical fiber based systems can be very useful in specific applications like the monitoring of road tunnels. The techniques already employed are all based on the deposition of a sensitive layer reacting with the gases, which change the optical properties of this layer (complex refractive index). Previously reported works used the deposition of a sol gel coating on a fiber core [3], on a fiber with a reduced diameter [4], on fiber Bragg gratings [5,6] or simply on the tip of a fiber [7,8]. Concerning the optical fiber sensors already developed for $NO_2$ measurement, one of the first configurations used sections of chemically pretreated porous silica fiber [9], which presented changes to the transmission spectra under $NO_2$ exposure. In [10], the sensor element consisted of the replacement of a portion of the cladding region of a multimode plastic clad silica fiber by metallophthalocyanines such as CuPc, PbPc and SmPc. There are some other works where the sensing element is extrinsic to the optical fiber. This is because the optical fiber is only used to transmit/receive the optical signal after the reflection in the sensitive material deposited in a sol-gel film [11] or a disc [12] or after transmission passing through a sensing plate [13].

In this work, the results obtained with an $NO_2$ optical fiber sensor based on a coated fiber tip, using LuPc2 as a sensitive molecule, are presented. The sensitive film experiences a variation of reflectance due to the decreased absorption at 660 nm following $NO_2$ adsorption. The adsorption is reversible but very slow at room temperature. So as to shorten the recovery time [12], the films were exposed to ultraviolet (UV) light at 365 nm. In order to get a practical system, both UV and red lights were injected into the same fiber, avoiding the need for an external UV source, and achieving a high power density of UV with a low power source thanks to the reduced size of the fiber core. This feature allows the simplification of the sensor head, making the external UV illumination of the film, reported in the works published up to now, unnecessary.

# 2. Material and Methods

## 2.1. Preparation of the Sensors

Phthalocyanines are good materials for gas sensing purposes due to their electrochemical and optical properties [14–16]. These organic molecules are known to be sensitive to oxidizing or reducing gases at ppm concentrations. For this reason they have been proposed as the chemically active component of both conductive and optical gas sensors [17–19]. Another major advantage is the remarkable chemical and thermal stability of the phthalocyanine derivatives in many environmental conditions. Lanthanide bisphthalocyanine (LnPc2) complexes with a "double-decker" structure [20], like that presented in

Figure 1 for LuPc2, are a typical class of compound with π-π* transitions and they exist in different forms associated with different colors. The green species is a neutral form, the red form is a singly oxidized species and the blue form is a singly reduced species. The study of lanthanide diphthalocyanines is very attractive due to their electrochromic properties [21–23] and semiconducting behaviors [24]. Among these lanthanide diphthalocyanines, the bisphtalocyanine of lutetium LuPc2 was chosen as the model compound for application in the present study.

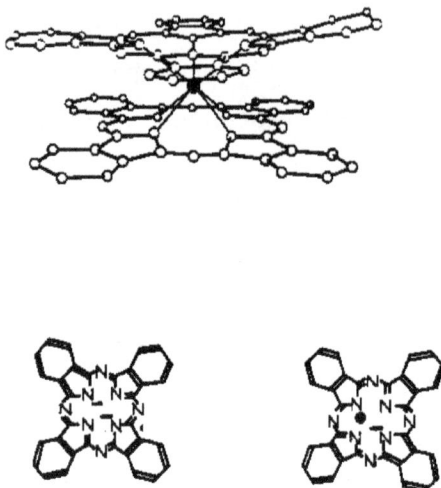

**Figure 1.** Molecular structure of lutetium bisphthalocyanine (LuPc2).

Low ionization energy together with high polarization energy makes the transfer of charge between LuPc2 and acceptor molecules easier. The transfer of electrons impacts the absorption spectra or the conductivity of LuPc2 in the UV, visible and near infrared ranges. These molecules were studied for gas detection [25–27].

The interaction of the LuPc2 molecules in the presence of NO2 gas is described in Equation (1). This reaction is reversible.

$$LuPc_2 + NO_2 \leftrightarrow LuPc_2^+ + NO_2^-$$ (1)

The sensor consists of a thin layer of LuPc2, deposited by evaporation on the tip of an optical fiber. The optical fiber is a pure silica 400 μm core diameter multimode fiber covered with a 12.5 μm thickness hard polymer as cladding. This optical fiber has a high OH content in order to reduce the optical transmission losses at UV and visible wavelengths.

Lutetium bisphthalocyanine (LuPc2) was synthetized by Bouvet (University of Burgundy, France) from the o-dicyanobenzene by heating it with lutetium triacetate, Lu(OAc)3, at approximately 300 °C, without any solvent according to [28].

The LuPc2 thin films were deposited in a BOC Edwards Auto 306 Evaporator by thermal sublimation at high vacuum (p = 2 × 10⁻⁶ mbar) with thicknesses ranging from 15 to 90 nm. The substrate was kept at room temperature. The deposition rate was ~5 nm/min and the film thickness was controlled *in situ* by using a quartz crystal thickness monitor located in the deposition chamber.

The LuPc2 deposits were carried out on the multimode optical fibers and on flat glass substrates in order to characterize the response to NO2 exposure on large samples.

## 2.2. Effect of NO2 on the Absorption Spectrum of LuPc2 and Accelerated Recovery by UV Light Illumination

Figure 2a,b show the absorbance spectra for 15 nm and 30 nm thicknesses of LuPc2 films on glass substrate, before and after contact with NO2. The film is originally green and fades to a reddish color after NO2 exposure. The spectrum is affected in the wavelength range between 400 nm and 1600 nm. Several absorption bands can be chosen to monitor optical changes. The most intense absorption band (Q band) at around 660 mm was chosen for this study. It can be seen that the strong absorption peak intensity decreases after only 1 min of NO2 exposure at a concentration of 100 ppm. This effect is reversible, but more than 24 h are needed at room temperature to recover the original spectrum. This is due to the high stability of the (LuPc2 + NO2$^-$) complex formed between NO2 and LuPc2, which induces a slow desorption of NO2.

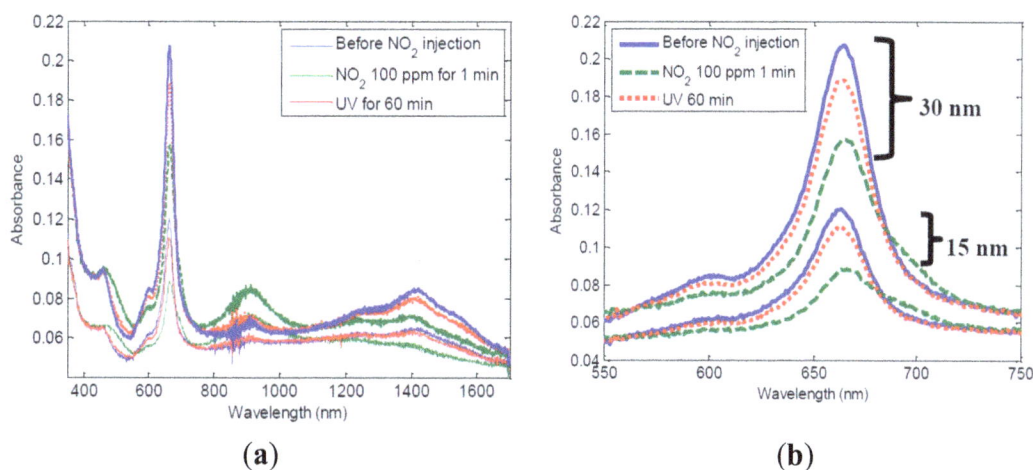

**Figure 2.** Absorbance spectra of pure LuPc2 thin films (**a**) whole spectra; (**b**) zoom on 660 nm.

However, it can be observed that UV illumination accelerates the recovery (the absorption peak increases after UV exposure) as shown by Baldini *et al* [12]. For this test, samples were placed in a 10 mW/cm$^2$ UV furnace at 365 nm for 60 min. This accelerated recovery is related to photodissociation of NO2 under UV light. The NO2 photodissociation under UV illumination at wavelengths below 420 nm can be expressed as:

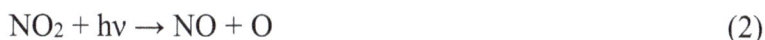

$$NO_2 + h\nu \rightarrow NO + O \tag{2}$$

The photodissociation rate for NO2 molecules under UV light illumination is proportional to the NO2 absorption cross-section and the quantum-yield for photodissociation [29], having a maximum rate in the range 360–395 nm. For the present experiments, a commercial LED source at 365 nm was used. It can be assured that prolonged exposures of the films at this wavelength do not destroy the films.

## 2.3. Measurement Setup

The experimental setup for NO2 measurement is shown in Figure 3. In the gas control part, a compressed air bottle was hooked up to two flow meters in order to control the relative humidity by changing the dry air/wet air relationship. A third flowmeter was connected to a 100 ppm NO2 bottle. The outputs from the three flowmeters were interconnected and attached to a glass cell where the optical fiber sensors were placed. In the output of the glass cell a $NO_x$ chemiluminescence analyzer was connected in order to measure the $NO_2$ concentration in the cell.

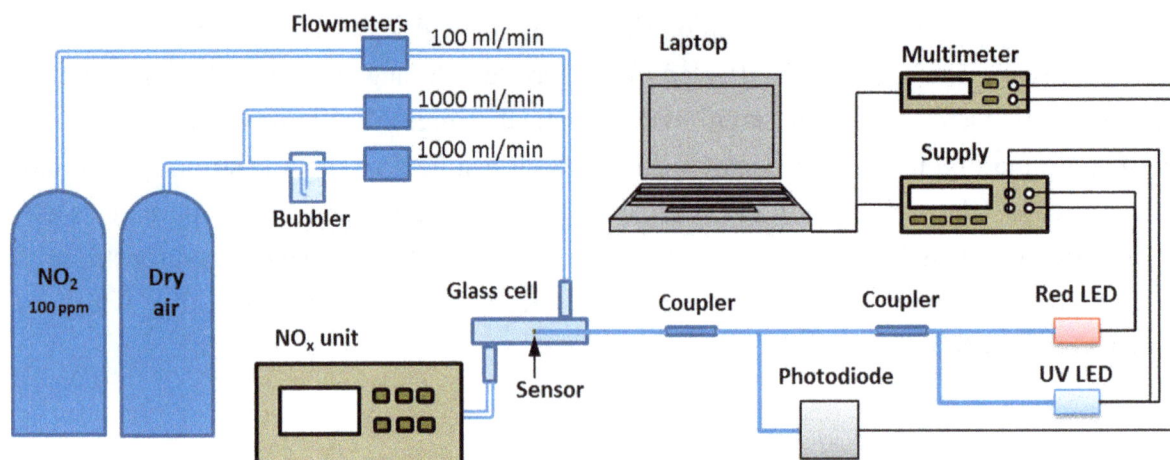

**Figure 3.** Experimental setup for $NO_2$ measurements.

In the optical part, two LEDs were used: a red one (emitting at 660 nm) and a UV one (emitting at 365 nm). A 200 μm core multimode optical fiber was connected to the UV LED, obtaining 0.5 mW of optical output power. A 400 μm core multimode optical fiber was connected to the red LED, yielding an optical output power of 14.5 mW. Both LEDs were connected to an optical coupler with a 50:50 ratio in order to combine the two wavelengths in the same fiber. The output of the coupler was connected to the input of an identical coupler. At the output of the second coupler a 400 μm core multimode fiber was connected, having the sensitive coating on the fiber end. Light was reflected by the fiber end, heading back to the optical coupler, and it was finally detected by a photodiode. The optical signal was converted into an electrical signal and it was measured with a multimeter. Because UV light disturbs the measurement by adding a superimposed voltage level, a laptop controlled the source to be synchronized with the multimeter. In order to take a measurement, the power supply of the UV LED was disabled and the voltage measured by the multimeter was registered. After that, the UV LED was switched on again. The power density at the end of the fiber was about 290 $W/m^2$.

## 3. Results and Discussion

### 3.1. Response of the Optical Fiber Sensor

The test presented in Figure 4 consisted of applying a flow of 4 ppm $NO_2$ for 15 min in humid air (50% RH) and after that, stopping the flow of $NO_2$ without applying UV illumination. As can be seen in Figure 4, the sensor reacted to the presence of $NO_2$ leading to an increase of the signal corresponding to an increasing reflectance of the sample. When the flow of $NO_2$ was stopped, the signal decreased very slowly since $NO_2$ desorption is very slow. The same experiment was then conducted with a simultaneous injection of UV light into the fiber. The adopted UV light illumination strategy was to activate the UV LED continuously, but since UV light disturbs the photodiode adding a superimposed voltage level, the

UV LED was disabled every 5 s in order to register the voltage for 1 s. Figure 4 shows that the use of UV-light leads to a drastic reduction of the recovery time.

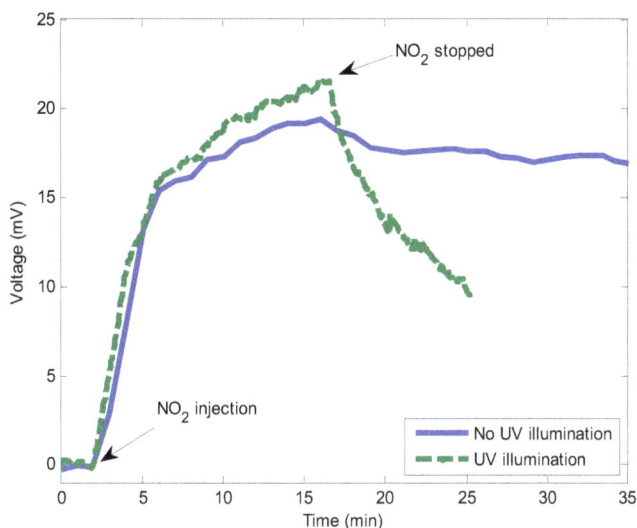

**Figure 4.** Response of a 15 nm thickness sensor without and with UV light illumination.

*3.2. Effect of the Thickness*

Up to three different coating thicknesses were deposited on 400 μm core multimode fiber tips: 15 nm, 30 nm and 45 nm. Optical sensors were placed inside the glass cell and a relative humidity of 50% was applied. Experiments started 1 h after signal stabilization.

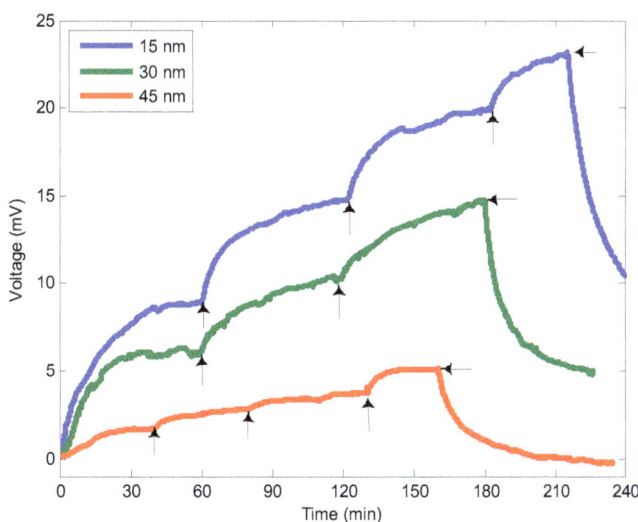

**Figure 5.** Response of the optical fiber sensors to different $NO_2$ concentrations.

Sensor responses to different $NO_2$ concentrations were measured. The $NO_2$ concentration applied to the optical sensors was increased in steps (25%, 50%, 75% and 100% of the $NO_2$ flowmeter capacity). These flow values correspond to measured $NO_2$ concentration values of 0.95 ppm, 2.00 ppm, 3.05 ppm and 4.10 ppm, respectively. The signal evolution from the different optical fiber sensors is shown in Figure 5. Sensors react to the increasing $NO_2$ concentrations by an increase of the signal on the photodiode. The injection times of NO$_2$ were not the same for the different sensors. These are marked in Figure 5 with arrows.

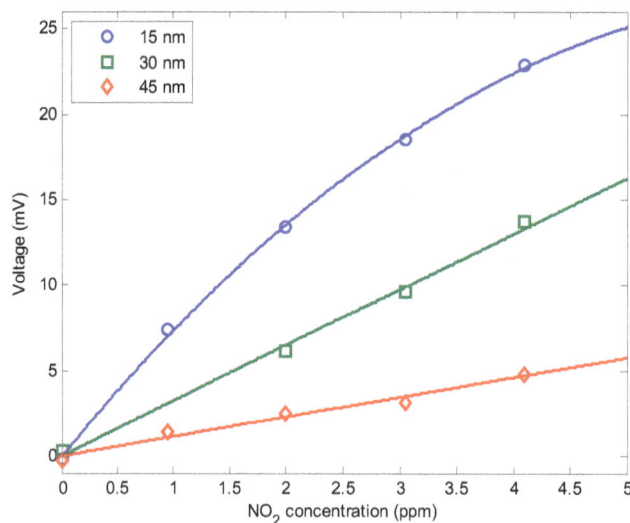

**Figure 6.** Calibration of the optical fiber sensors with NO$_2$ concentration.

Figure 6 depicts the response after stabilization versus NO$_2$ concentration for different thicknesses. The signal decreases when the thickness increases, which means that the reaction remains close to the film surface. Moreover, the response time was not dramatically increased by an increase of the thickness (see Table 1) confirming that gas does not diffuse throughout the film. If this were the case, the response time would be multiplied by a factor four when doubling the thickness according to the classical Fick diffusion equations.

The response of the 15 nm thickness sensor shows a non-linear response, probably due to a saturation mechanism, whereas the 30 nm and the 45 nm can be linearly fitted. The sensitivities of the sensors with coating thicknesses of 15 nm, 30 nm and 45 nm can be calculated as $S_{15nm}$ = 5.02 mV/ppm, $S_{30nm}$ = 3.25 mV/ppm and $S_{45nm}$ = 1.15 mV/ppm, respectively. Using a digital multimeter with a display resolution of 6½ digits, a voltage resolution of 0.001 mV can be obtained, corresponding to an NO$_2$ concentration resolution of 0.2 ppb, 0.3 ppb and 0.9 ppb for thicknesses of 15 nm, 30 nm and 45 nm, respectively.

*3.3. Repeatability*

Repeatability tests were also carried out. The response of the 15 nm thickness sensor to consecutive injections of 2 ppm NO$_2$ is depicted in Figure 7. As it can be seen, the response is quite repeatable. The standard deviation can be calculated as 5.54%. The repeatability performances for the 30 nm and 45 nm thicknesses sensors are similar. Concerning the 15 nm thickness sensor, the mean response time (defined as the time needed to reach the 90% of the final voltage) was 25 min and the mean recovery time (defined as the time needed to reach the 10% of the final voltage) was 38 min. Concerning the 30 nm thickness sensor, the response time was 27 min and the recovery time was 35 min. Finally, for the 45 nm thickness sensor, the response time was 27 min and the recovery time was 40 min.

The response or recovery time does not drastically depend on the thickness, confirming that NO2 does not diffuse completely in the film.

To summarize, the relationship between the detected voltage and the NO2 concentration is given in Table 1.

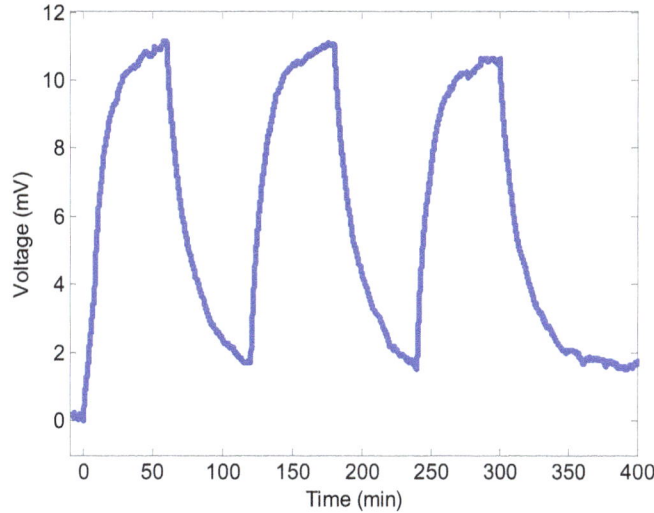

**Figure 7.** Repeatability performance for the 15 nm thickness NO2 optical fiber sensor.

**Table 1.** Characterization of the sensors depending on their thicknesses.

|        | Sensibility | Resolution | Response time | Recovery Time |
|--------|-------------|------------|---------------|---------------|
| 15 nm  | 5.02 mV/ppm | 0.2 ppb    | 25 min        | 38 min        |
| 30 nm  | 3.25 mV/ppm | 0.3 ppb    | 27 min        | 35 min        |
| 45 nm  | 1.15 mV/ppm | 0.9 ppb    | 27 min        | 40 min        |

Compared to a previously developed NO2 optical fiber sensors [12] working in a similar range of NO2 concentrations (between 0 ppm and 10 ppm), the sensor presented in this article has lower response and recovery times, but most importantly, the UV illumination for signal recovery can be done without direct intervention. UV light can be transmitted through the fiber and there is no need to disconnect the fiber in order to illuminate the sensitive material with an external UV lamp. In a more recent work [13], the sensor presented worked in a higher NO2 concentration range (between 0 ppm and 30 ppm), with a similar response and recovery time. Nevertheless, the recovery time can be decreased, but only by applying heat and a vacuum to the sensitive material. It has to be noted that the two aforementioned sensors are extrinsic to the optical fiber, since the sensitive material is coated in an external support (a membrane or a plate), making the sensor head more complex and bigger in size.

*3.4. Sensor Modeling*

In this section, the equations describing the behavior of the optical fiber NO2 sensor are developed. The measurand related to the environment (NO2 concentration) in the setup presented in Figure 3 is the photodetector voltage as a result of the photodetected optical power reflected from the sensor. The photodetector voltage can be calculated with Equation (3):

$$V_{PD} = P_{in} \cdot \Re_{PD} \cdot G_{PD} \tag{3}$$

where $P_{in}$ is the input power to the photodetector in W, $\Re_{PD}$ is the responsivity of the photodiode in A/W, and $G_{PD}$ is the transimpedance gain of the photodiode in V/A. The input power to the photodiode can be calculated with following equation:

$$P_{in} = \frac{P_{LED}}{L_T} \cdot R = P_{LED} \cdot G_T \cdot R \tag{4}$$

where $P_{LED}$ is the output power of the red LED (emitting at 660 nm) in W, $L_T$ is the total power loss of the optical power in the optical setup (including connector losses and power splitting and coupling efficiency in the optical couplers) and R is the combined reflectance of the fiber and the sensitive layer. The optical losses can be considered as a gain as the $G_T$ value is lower than 1. By incorporating Equation (4) into Equation (3), it can be seen that the photodetector voltage depends on the reflectance, since the other parameters are constant:

$$V_{PD} = P_{LED} \cdot G_T \cdot \Re_{PD} \cdot G_{PD} \cdot R = K \cdot R \tag{5}$$

The reflectance of the combination of the fiber and sensitive layer can be described as the sum of the reflectance in the fiber-layer interface ($R_{fl}$) and the reflectance in the layer-air interface ($R_{la}$):

$$R \approx R_{fl} + R_{la} \cdot e^{-2\alpha d} \tag{6}$$

The light reflected in the layer-air interface experiences attenuation due to the absorption of the sensitive layer of length d, characterized by an absorption coefficient $\alpha$, expressed in $m^{-1}$. This coefficient is homogeneous in the sensitive layer but after $NO_2$ exposure there is a diffusion of the gas through the layer, decreasing the absorbance (as seen in Figure 2) in a length $d_g$, as is depicted in Figure 8b. Thus, after $NO_2$ exposure the total absorption in the layer can be described as follows:

$$\alpha d = \alpha_n d_n + \alpha_g d_g \tag{7}$$

where $\alpha_n$ is the absorption coefficient of the sensitive layer in a normal situation (before $NO_2$ exposure) in the unaffected length $d_n$ and $\alpha_g$ is the absorption coefficient in the length $d_g$ where the $NO_2$ gas penetrates.

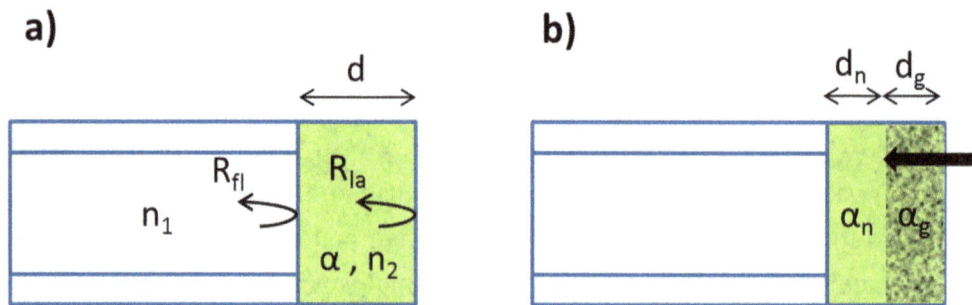

**Figure 8.** Sensor representation (**a**) before and (**b**) after $NO_2$ exposure.

Inserting Equations (7) and (6) into Equation (5), the photodetector voltage can be expressed as:

$$V_{PD} = K[R_{fl} + R_{la} \cdot e^{-2\alpha d}] = K[R_{fl} + R_{la} \cdot e^{-2(\alpha_n d_n + \alpha_g d_g)}] = K[R_{fl} + R_{la} \cdot e^{-2(\alpha_n(d-d_g)+\alpha_g d_g)}]$$

$$= K[R_{fl} + R_{la} \cdot e^{-2\alpha_n d} \cdot e^{-2(\alpha_g-\alpha_n)d_g}] \approx K[R_{fl} + R_{la} \cdot e^{-2\alpha_n d}(1 + 2(\alpha_n - \alpha_g)d_g)]$$

$$V_{PD} = K[R_{fl} + R_{la} \cdot e^{-2\alpha_n d}] + 2K[R_{la} \cdot e^{-2\alpha_n d}(\alpha_n - \alpha_g)d_g] = V_0 + \Delta V_{PD} \tag{8}$$

In Equation (8), the first term $V_0$ is a baseline and the voltage variations are encoded in the second term.

$$\Delta V_{PD} = 2KR_{la} \cdot e^{-2\alpha_n d}(\alpha_n - \alpha_g)d_g \qquad (9)$$

The variation of NO2 concentration modifies the parameters $\alpha_g$ and $d_g$ since the penetration of the NO2 molecules and the absorption depend on the thickness of the layer and the NO2 concentration as can be seen in Figure 2. According to Equation (9), and due to the fact that the gas does not diffuse in the whole layer, the voltage variation $\Delta V_{PD}$ depends on the thickness of the layer. When d increases the sensitivity decreases, as was experimentally observed.

## 4. Conclusions

An optical fiber chemical sensor based on LuPc2 thin films has been developed for NO2 detection, which is useful for the monitoring of traffic pollution. A special fiber was selected in order to allow the transmission of both UV and visible light through the fiber, improving the previously developed NO2 sensors by the simplification of the sensor head, since there is no need to use an external UV source to illuminate the films.

The sensitive coating was deposited by thermal evaporation onto the fiber tip. Absorption at 660 nm was monitored. Contact with NO2 drastically and reversibly decreased absorption. Due to a strong interaction with NO2, the LuPc2 thin films showed good sensitivity for low concentrations, but the recovery time was very long. To reduce the recovery time, the films were irradiated by UV light at 365 nm in order to promote the photodissociation of the complex formed between NO2 and LuPc2 molecules.

In this study, the sensors demonstrated the possibility of monitoring NO2 concentration in the range 0–5 ppm. Sensitivity decreases when the film thickness increases, but the impact of the thickness on the response or recovery time is less important. This means that the reaction essentially takes place close to the surface. A higher sensitivity of 5.02 mV/ppm was obtained with the 15 nm coating thickness sensor, also achieving lower response and recovery times of 25 min and 38 min, respectively. A resolution of 0.2 ppb was obtained with this sensor.

## Acknowledgments

This work was supported by the FEDER program (E.U.) and by the GREENOMAT and WBGREEN programs (Walloon Region of Belgium). The authors also acknowledge the financial support of ERC (the Starting Independent Researcher Grant "PROSPER"). C. Caucheteur is supported by the F.R.S.-FNRS. The authors would like to thank Marcel Bouvet from the University of Burgundy (France) for collaborating in the preparation of a synthesis of LuPc2 in powder.

## Author Contributions

Antonio Bueno performed the experiments with the optical fibers and wrote the paper. He also selected and acquired the components for the experiments and made the setup. Driss Lahem contributed to the synthesis and purification of the LuPc2 molecule and made the structural and functional characterization experiments of this molecule in thin films. Marc Debliquy had the original idea of the simultaneous transmission of UV and visible light and he conceived the experiments together with Christophe Caucheteur. All the authors have actively contributed to the review process of the paper.

**Conflicts of Interest**

The authors declare no conflict of interest.

**References**

1.  Olivier, J.J.M.; Bouwman, J.G.J.; van der Hoek, A.F.; Berdowski, K.W. Global air emission inventories for anthropogenic sources of $NO_x$, $NH_3$ and $N_2O$ in 1990. *Environ. Pollut.* **1998**, *102*, 135–148.

2.  Maître, A.; Bonneterre, V.; Huillard, L.; Sabatier, P.; Gaudemaris, R. Impact of urban atmospheric pollution on coronary disease. *Eur. Heart J.* **2006**, *27*, 2275–2284.

3.  Mac Craith, B.D.; O'Keefe, G.; McDonagh, C.; McEvoy, A.K. LED-based fibre optic oxygen sensor using sol-gel coating. *Electron. Lett.* **1994**, *30*, 888–889.

4.  Bariáin, C.; Matías, I.R.; Arregui, F.J.; López-Amo, M. Experimental results towards development of humidity sensors by using hygroscopic material on biconically tapered optical fiber. *Proc. SPIE* **1998**, *3555*, 95–105.

5.  Caucheteur, C.; Debliquy, M.; Lahem, D.; Mégret, P. Catalytic fiber Bragg grating sensor for hydrogen leak detection in air. *IEEE Photon. Technol. Lett.* **2008**, *20*, 96–98.

6.  Caucheteur, C.; Debliquy, M.; Lahem, D.; Mégret, P. Hybrid fiber gratings coated with a catalytic sensitive layer for hydrogen sensing in air. *Opt. Express* **2008**, *16*, 16854–16859.

7.  Mechery, S.J.; Singh, J.P. Fiber optic based gas sensor with nanoporous structure for the selective detection of $NO_2$ in air samples. *Anal. Chimica Acta* **2006**, *557*, 123–129.

8.  Bezunartea, M.; Estella, J.; Echeverría, J.C.; Elosúa, C.; Bariáin, C.; Laguna, M.; Luquin, A.; Garrido, J.J. Optical fibre sensing element based on xerogel-supported $[Au_2Ag_2(C_6F_5)_4(C_{14}H_{10})]n$ for the detection of methanol and ethanol in the vapour phase. *Sens. Actuators B Chem.* **2008**, *134*, 966–973.

9.  Schmidlin, E.M.; Mendoza, E.A.; Ferrell, D.J.; Syracuse, S.J.; Khalil, A.N.; Lieberman, R.A. A fiber optic $NO_2$ sensor for combustion monitoring. *Proc. SPIE* **1994**, *2068*, 41–48.

10.  John, M.S.; Unnikrishnan, K.P.; Thomas, J.; Radhakrishnan, P.; Nampori, V.P.N.; Vallabhan, C.P.G. Characterization of an optical fiber sensor in detecting $NO_2$ gas. In Proceedings of the International Conference on Fiber Optics & Photonics PHOTONICS-2000, Kolkata, India, 18–20 December 2000; pp. 612–616.

11.  Mechery, S.J.; Singh, J.P. Self-calibrated fiber optic transflection probe for $NO_2$ detection. *Proc. SPIE* **2004**, *5272*, 110–115.

12.  Baldini, F.; Capobianchi, A.; Falai, A.; Mencaglia, A.A.; Pennesi, G. Reversible and selective detection of $NO_2$ by means of optical fiber. *Sens. Actuators B Chem.* **2001**, *74*, 12–17.

13.  Ohira, S.-I.; Wanigasekara, E.; Rudkevich, D.M.; Dasgupta, P.K. Sensing parts per million levels of gaseous $NO_2$ by an optical fiber transducer based on calix[4]arenes. *Talanta* **2009**, *77*, 1814–1820.

14. Leznoff, C.C.; Lever, A.B.P. *Phthalocyanines: Properties and Applications*; VCH Publishers: New York, NY, USA, 1989; Vol. 1–3.

15. McKeown, N.B. *Phthalocyanine Materials: Synthesis, Structure and Function. Chemistry of Solid State Materials*; Cambridge University Press: Cambridge, UK, 1998; Vol. 6.

16. Simon, J.; André, J.-J. *Molecular Semiconductors*. Springer Verlag Berlin Heidelberg: Berlin, Germany, 1985.

17. Wright, J.D. Gas adsorption on phthalocyanines and its effects on electrical properties. *Prog. Surf. Sci.* **1991**, *31*, 1–60.

18. Mukhopadhyay, S.; Hogarth, C.A. Gas sensing properties of phthalocyanine Langmuir–Blodgett films. *Adv. Mater.* **1994**, *6*, 162–164.

19. Capone, S.; Mongelli, S.; Rella, R.; Siciliano, P.; Valli, L. Gas sensitivity measurements on NO2 sensors based on Coper(II) tetrakis(n-butylaminocarbonyl) phthalocyanine LB films. *Langmuir* **1999**, *15*, 1748–1753.

20. Simon, J.; Bouvet, M.; Bassoul, P. *The Encyclopedia of Advanced Materials*; Pergamon Press: Oxford, UK, 1994; pp. 1680–1692.

21. Rodríguez-Méndez, M.L.; Gorbunova, Y.; de Saja, J.A. Spectroscopic Properties of Langmuir−Blodgett Films of Lanthanide Bis(phthalocyanine)s Exposed to Volatile Organic Compounds. Sensing Applications. *Langmuir* **2002**, *18*, 9560–9565.

22. Rodriguez-Mendez, M.L.; Aroca, R.; DeSaja, J.A. Electrochromic and gas adsorption properties of Langmuir-Blodgett films of lutetium bisphthalocyanine complexes. *Chem. Mater.* **1993**, 5, 933–937.

23. Rodriguez-Mendez, M.L.; Aroca, R.; DeSaja, J.A. Electrochromic properties of Langmuir-Blodgett films of bisphthalocyanine complexes of rare earth elements. *Chem. Mater.* **1992**, *4*, 1017–1020.

24. Maitrot, M.; Guillaud, G.; Boudjema, B.; André, J.-J.; Strzelecka, H.; Simon, J.; Even, R. Lutetium bisphthalocyanine: The first molecular semiconductor. Conduction properties of thin films of p- and n-doped materials. *Chem. Phys. Lett.* **1987**, *133*, 59–62.

25. Gutierrez, N.; Rodríguez-Méndez, M.L.; de Saja, J.A. Array of sensors based on lanthanide bisphtahlocyanine Langmuir–Blodgett films for the detection of olive oil aroma. *Sens. Actuators B Chem.* **2001**, *77*, 437–442.

26. De Saja, J.A.; Rodríguez-Méndez, M.L. Sensors based on double-decker rare earth phthalocyanines. *Adv. Colloid Interf. Sci.* **2005**, *116*, 1–11.

27. Bariáin, C.; Matías, I.R.; Fernández-Valdivielso, C.; Arregui, F.J.; Rodríguez-Méndez, M.L.; de Saja, J.A. Optical fiber sensor based on lutetium bisphthalocyanine for the detection of gases using standard telecommunication wavelengths. *Sens. Actuators B Chem.* **2003**, *93*, 153–158.

28. Clarisse, C.; Riou, M.-T. Synthesis and characterization of some lanthanide phthalocyanines. *Inorg. Chimica Acta* **1987**, *130*, 139–144.

29. Parrish, D.D.; Murphy, P.C.; Albritton, D.L.; Fehsenfeld, F.C. The measurement of the photodissociation rate of NO2 in the atmosphere. *Atmos. Environ.* **1983**, *17*, 1365–1379.

# A Low-Cost Sensing System for Cooperative Air Quality Monitoring in Urban Areas

**Simone Brienza** [1,*]**, Andrea Galli** [1]**, Giuseppe Anastasi** [1] **and Paolo Bruschi** [2]

[1] Department of Information Engineering, University of Pisa, Largo Lucio Lazzarino 1, 56122 Pisa, Italy; E-Mails: galli@andrea.si (A.G.); giuseppe.anastasi@unipi.it (G.A.)

[2] Department of Information Engineering, University of Pisa, Via G. Caruso 16, 56122 Pisa, Italy; E-Mail: paolo.bruschi@unipi.it

\* Author to whom correspondence should be addressed; E-Mail: simone.brienza@for.unipi.it

Academic Editor: Antonio Puliafito

**Abstract:** Air quality in urban areas is a very important topic as it closely affects the health of citizens. Recent studies highlight that the exposure to polluted air can increase the incidence of diseases and deteriorate the quality of life. Hence, it is necessary to develop tools for real-time air quality monitoring, so as to allow appropriate and timely decisions. In this paper, we present *uSense*, a low-cost cooperative monitoring tool that allows knowing, in real-time, the concentrations of polluting gases in various areas of the city. Specifically, users monitor the areas of their interest by deploying low-cost and low-power sensor nodes. In addition, they can share the collected data following a social networking approach. *uSense* has been tested through an in-field experimentation performed in different areas of a city. The obtained results are in line with those provided by the local environmental control authority and show that *uSense* can be profitably used for air quality monitoring.

**Keywords:** urban sensing; air quality monitoring; cooperative system; participatory sensing

## 1. Introduction

Air quality is a major concern for the public health, the environment and, ultimately, the economy of all the industrialized countries. In the last years, Europe and USA have considerably reduced the emissions of several airborne pollutants [1,2] such as carbon monoxide (CO), benzene ($C_6H_6$), sulphur dioxide ($SO_2$), and lead (Pb). Anyway, nitrogen dioxide ($NO_2$), ozone ($O_3$), particulate matter (PM), and some other organic compounds still represent a serious threat. In such context, the "Air Quality in Europe" report—published by the European Environment Agency (EEA) [3] in November 2014—provides a thorough overview of measures and policies adopted at European level to improve the air quality and reduce the impact of air pollution on public health and ecosystems. In addition, the report describes the effects of air pollution on health, climate, and ecosystems. As shown also in [4], poor air quality can, in fact, cause ill health and premature deaths, as well as damages to ecosystems, crops, and buildings. Obviously, the effects are more serious in urban areas, where the majority of the population lives.

For all these reasons, an accurate real-time monitoring of air quality in urban areas is essential to enable appropriate and timely public decisions and, ultimately, is extremely important for preserving the citizens' health. However, it is often difficult for citizens to obtain pollution data. In fact, air quality is usually monitored through large and expensive sensing stations, installed at some strategic locations, and managed by public authorities. Therefore, the monitoring (although accurate) is limited to few specific areas and the gathered measurements are sometimes not available to citizens. On the other hand, people are very interested in knowing air quality conditions in places where they live and spend much of their time, such as the area where their home is located, the school of their kids, their working place, public gardens, *etc*.

To overcome these limitations, in this paper, we present *uSense*, a sensing system for cooperative air quality monitoring in urban areas. *uSense* relies on small-size low-cost sensor nodes (equipped with gas sensors such as $O_3$, CO and $NO_2$) that can be privately installed by citizens inside their properties. Sensor nodes are powered by long-lasting batteries and use WiFi for data transfer, which makes their deployment very flexible and allows to easily place them outdoors. Owners of sensor nodes can also share their measurements, through a social networking approach, thus enabling cooperative sensing. This allows a real-time and fine-grained air quality monitoring of urban areas. The proposed system has been implemented and tested through in-field experimentation.

The main goal of our study is to investigate whether a low-cost monitoring system can provide reliable indications about air quality in a place and, hence, can be used in practice. To this end, we have compared the pollution measurements obtained by *uSense* with the limit values specified by law and with the official measurements made available by the local control authority. From our experimental study, it emerges that, even though measurements provided by low-cost sensors are not as accurate as official data, nevertheless they can provide useful information about air quality in a specific location.

The rest of the paper is organized as follows. Section 2 describes the previous solutions already present in the literature, remarking the improvements introduced by *uSense*. Section 3 provides an overall description of *uSense*. Section 4 introduces the concept of Air Quality Index and shows how it can be calculated. Section 5 discusses the system architecture, whereas Section 6 provides some implementation details. Section 7 shows the results obtained using *uSense* to monitor some urban areas for one month (May 2014). Finally, Section 8 concludes the paper.

## 2. Previous Solutions

Sharing data and information with other people is becoming more and more common, thanks to the increasing popularity of personal devices (e.g., smartphones) that integrate low-cost sensors and provide fast and reliable data connectivity. Following an approach known as *Participatory Sensing*, users can gather information of interest and make it accessible to other people. Generally, data shared by users—relative to their positions—are processed and aggregated. This way, the other users of the system can access more complete and reliable information. Due to this trend, in recent years we witnessed a proliferation of devices and applications aimed at collecting and sharing data. The areas of application are many, ranging from health and fitness to environmental monitoring, from transportation monitoring to urban sensing. Especially in urban areas, many solutions have been proposed in order to improve the quality of life.

For instance, in [5] the authors propose *Ear-Phone*, a system that makes use of the microphone mounted on smartphones (preliminarily calibrated) to produce noise maps of a city, so as to allow citizens to avoid the noisiest areas.

Similarly, several solutions have been proposed in the literature in order to monitor airborne pollutants with mobile low-cost sensors and share the obtained measurements with other users. For instance, *InAir* [6] and *MAQS* [7] face the issue of indoor air quality sensing. Specifically, *InAir* relies on stationary gas sensors placed inside users' homes. A dedicated display on each sensor node visualizes the air quality values measured in the room and the values acquired by the other nodes in the other parts of the house. *MAQS*, instead, considers a mobile wearable sensing system that provides air quality information for each room visited by the user. The obtained data can be shared, through smartphones, with people who do not carry the sensing system.

However, our interest is mainly focused on outdoor sensing. Also in this case, wearable sensors have been used to monitor air quality, for instance in *CitiSense* [8] and *Common Sense* [9]. Both these solutions rely on small, battery-powered sensor nodes that measure the concentrations of polluting gases and send air quality data to users' smartphones through Bluetooth. The obtained data and the GPS coordinates are then shared with other users through a dedicated website, an Android app or social networks. Obviously, the small size of the sensor nodes and the required connectivity highly affect the lifetime of the battery, which needs to be frequently recharged.

A similar approach is also followed in [10] where the authors propose *GasMobile*, a small and portable ozone measurement system. Essentially, it relies on a particular sensor equipped with a serial transmitter board that allows to directly connect the sensor to the user's smartphone through a USB port. The system is extremely compact and leverages an Android application to calibrate the sensor and upload the data to an ad hoc server. However, the readings are very limited, since they refer just to a single gas. Extending the measurements to other substances would considerably complicate the device, increasing its size and shortening its lifetime.

Conversely, in [11] the authors propose a solution to derive high-resolution air pollution maps for urban areas, using nodes provided with several sensors, such as UFP (ultrafine particles), CO, $O_3$, and $NO_2$ sensors. Basically, the node acquires a location information through its GPS receiver, then, it transmits the gathered pollution measurements via the cellular network (*i.e.*, GSM) to the back-end server for further processing. The authors also propose a novel modeling approach to create pollution maps with high spatial and temporal resolution starting from the obtained measurements. However, differently from *uSense*, nodes are installed on buses and cannot be moved, once deployed. Therefore, they cannot be directly managed by citizens. This is a major drawback, since a user could not be able to know the pollution level in the areas of her/his interest.

Compared to the previously presented solutions, *uSense* allows to monitor the concentration levels of several air pollutants (*i.e.*, Ozone $O_3$, Carbon monoxide CO, and Nitrogen dioxide $NO_2$), by using small low-cost sensor nodes. These nodes can be deployed by the users in their areas of interest and moved whenever they want to monitor a different place. In fact, the proposed solution leverages on long-lasting batteries, which do not need to be frequently recharged, and exploits WiFi for data transfer, so that nodes can be easily placed outside as well. The *uSense* solution presented in this paper extends a previous version proposed in [12]. With respect to it, a new formulation of the Air Quality Index has been considered (detailed in Section 4). Consequently, all the obtained results have been revised according to the new definition. In addition, the results have been compared with the official measurements published by the local environmental control agency.

## 3. System Overview

Through *uSense*, a user can monitor the air quality near her/his house, just by placing a small sensor node in her/his property, for instance in a garden, a balcony, a window sill, or hung to an outside wall. Basically, the node periodically measures—through proper sensors—the concentrations of some airborne pollutants. Then, the obtained data are sent to the *uSense* database via the Internet, and are made accessible to all the other *uSense* users. This way, in a cooperative fashion, each user contributes to monitor part of the city. Clearly, the system is particularly useful in urban areas where many sensor nodes have been placed. In this case, in fact—since pollution data are shared—all the *uSense* users can know, in real time, the pollution level of the various areas of the city. In addition, they can focus on specific regions of interest (e.g., their homes, workplaces, kids' schools, *etc.*) to visualize the pollution level in the current day, or in the past. Such information can be exploited by users in several ways. For instance, they can decide to reach a destination searching for the less polluted path, or avoid going out during the most polluted hours.

In details, the actions performed by *uSense*, from the initial acquisition of data to their publication, are the following:

(1) *Gas Sampling.* In order to monitor the air quality inside an area of interest, a user has only to deploy one or more sensor nodes. These nodes measure the concentration in the air of some damaging pollutants, e.g., *$O_3$ (Ozone), CO (Carbon monoxide),* and *$NO_2$ (Nitrogen dioxide).* The measurements are performed periodically (e.g., every 30 min).

(2) *Data Transfer.* After that the concentrations of the various pollutants have been measured, the obtained data are arranged in a packet and transmitted to the *uSense* server, through the Internet, using a wireless technology (e.g., WiFi or GPRS).

(3) *Air Quality Index Calculation.* Once the *uSense* server has received the gas concentrations from a sensor node, it calculates the *Air Quality Index (AQI)* for the area corresponding to the node. Essentially, the server derives a number that indicates how good/bad the air quality is, so that increasing values correspond to higher pollution levels. Formulas for AQI calculation are defined

by government agencies and vary from country to country and, sometimes, also from region to region, inside the same country. For instance, the index used in the USA—defined by the Environmental Protection Agency—is presented in [13]. In Europe, a Common Air Quality Index [14] has been developed, so as to present the air quality situation in European cities in a comparable and easily understandable way. However, the adoption of this index is not compulsory for European countries. Therefore, at the present day, only few cities and regions use it, while each country maintains its own regulations and indices. Details about the *Air Quality Index* adopted in *uSense* are provided in the following section.

(4) *Data sharing and view.* Data obtained from sensor nodes are stored inside a database and are made accessible to the *uSense* community, following a collaborative paradigm. This way, each user can view the AQI and gas levels regarding her/his nodes as well as the other users' nodes. In addition, users can simply access air quality data even if they do not participate in the monitoring process (*i.e.*, they do not own sensor nodes). *uSense* provides several mechanisms for exporting and visualizing data. They can be accessed through a Web interface, a mobile application, or Web services and can be presented in many ways. For instance, users can view the position of sensor nodes in a map, they can view the pollution levels of the various areas of the city, or search for the less polluted paths, and so on. Further details are provided in Section 5.

In order to be practically used by users, *uSense* provides some additional features:

— *Easy placement.* Sensor nodes are small and can be easily moved from a place to another. Furthermore, they can be placed outdoors, since they are battery-powered and the data communication is wireless.

— *Easy installation and access. uSense* can be easily set up by users. In fact, sensor nodes require just few steps to be fully working (see below for details). In addition, each user can access the data produced by her/his sensor nodes (or by the other users' ones), simply creating a personal account in *uSense*, through the Web interface.

— *Robustness.* Each sensor node is inserted in a box, aimed at protecting it from atmospheric agents and impacts. Moreover, in order to provide a reliable service, whenever the access point/router is temporarily not available (e.g., due to weather conditions, channel errors, connection problems, hardware crashes, or blackouts) data are provisionally stored in a SD card, until the wireless connection is available again.

## 4. Air Quality Index

In each country, air quality monitoring is regulated by law. For instance, in Europe, the air quality Directive 2008/50/EC [15] defines the limit values for each pollutant and specifies the methods for measuring pollutants' concentrations. Obviously, low-cost gas sensors (as the ones used in *uSense*) do not always reach the quality objectives prescribed by the legislation. Indeed, *uSense* is not intended to provide official measurements. Instead, it aims to provide useful and easy-to-read indications about the air quality in specific locations, making people aware of air pollution in the area where they live.

In this perspective, we have defined a simple Air Quality Index in order to provide an easily readable and understandable measure of air pollution level, even for non-experts. Basically, for each sensor node, *uSense* returns a number that synthetically represents the degree of pollution in the air. No specific knowledge is required to interpret this value. Simply, the higher the index value, the more the air is polluted. To calculate AQI, concentrations of several air pollutants are taken into account and compared to the limits fixed by law.

In *uSense*, the AQI is obtained by comparing the average concentrations of the monitored gases, measured in the considered day, to the limit values specified by national regulations, as follows:

$$AQI = \max\left\{\frac{G_1^{meas}}{G_1^{lim}}, \frac{G_2^{meas}}{G_2^{lim}}, \frac{G_3^{meas}}{G_3^{lim}}, \dots, \frac{G_N^{meas}}{G_N^{lim}}\right\} \tag{1}$$

Specifically, let us assume we are monitoring $N$ polluting gases. For each considered gas $G_i$ (with $i$ ranging from 1 to $N$), $G_i^{meas}$ is obtained from the measurements performed throughout the day, considering the averaging period specified by law, whereas $G_i^{lim}$ indicates the allowed concentration limit. In Equation (1), the concentration of each considered pollutant is divided for its reference limit. Then, the AQI is chosen as the highest value (corresponding to the pollutant with the highest concentration) among the obtained ratios. This way, the index warns us if one or more pollutants have exceeded the maximum concentration permitted.

In the current implementation of *uSense* we only monitor $O_3$, $NO_2$, and CO. Consequently, Equation (1) can be rewritten as:

$$AQI = \max\left\{\frac{O_3^{meas}}{O_3^{lim}}, \frac{NO_2^{meas}}{NO_2^{lim}}, \frac{CO^{meas}}{CO^{lim}}\right\} \tag{2}$$

Limit values and averaging periods for $O_3$, $NO_2$, and CO (as specified in [15]) are reported in Table 1.

**Table 1.** $O_3$, $NO_2$ and CO averaging periods and limit values.

| Pollutant | Averaging Period | Limit |
|-----------|------------------|-------|
| Ozone $O_3$ | maximum hourly average | 180 $\mu g/m^3$ |
| Nitrogen dioxide $NO_2$ | maximum hourly average | 200 $\mu g/m^3$ |
| Carbon monoxide CO | daily maximum 8-hour average | 10 $mg/m^3$ |

To make the AQI value easily interpretable by eye, we have considered five air quality classes and we have assigned a color to each one. The first two classes indicate that reference limits have not been exceeded and, thus, there are no problems related to air quality. The other three classes, instead, warn that—with different levels of severity—some gases have exceeded the limit fixed by law. Classes and colors are reported in Table 2.

**Table 2.** Air Quality Index and Levels.

| Air Quality Index | Air Quality Classes | Color |
|-------------------|---------------------|-------|
| From 0 to 0.5 | Good | |
| From 0.5 to 1 | Fair | |
| From 1 to 1.5 | Moderate | |
| From 1.5 to 2 | Unhealthy | |
| More than 2 | Insalubrious | |

## 5. System Architecture

In this section, we present the architecture of *uSense*. As shown in Figure 1, *uSense* includes several components, *i.e.*, *sensor nodes*, *wireless access points*, *server* and *end-user devices*. They are described in the following subsections.

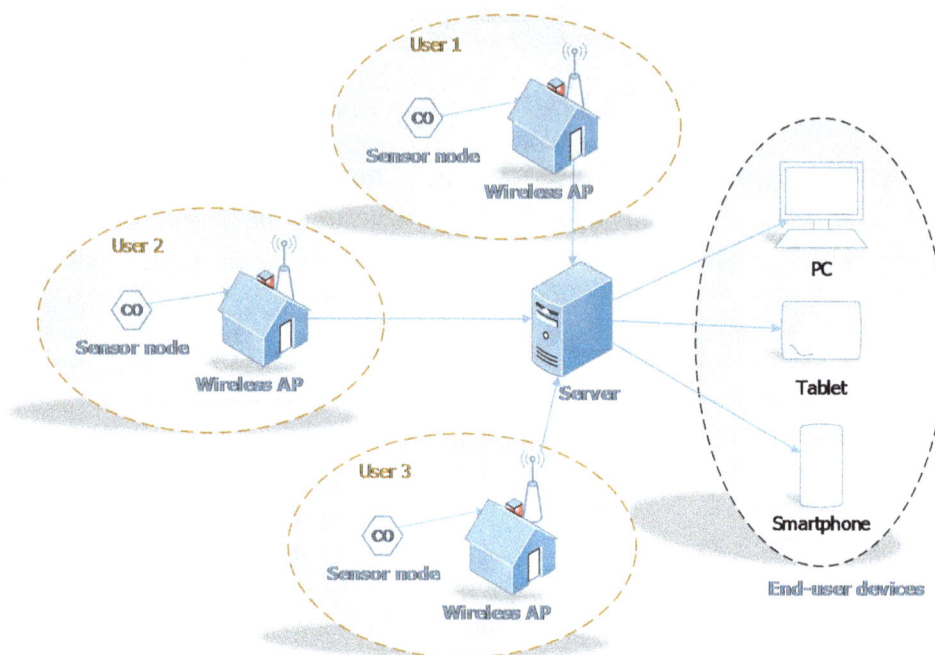

**Figure 1.** System architecture.

### 5.1. Sensor Nodes

*uSense* handles the measurements performed by a number of sensor nodes positioned at the users' premises. Basically, sensor nodes measure the concentration of the airborne pollutants and send the acquired data to the *uSense* server. In order to extend their battery life, they use a duty-cycle mechanism that alternates activity and sleep periods. Specifically, each sensor node periodically (e.g., every 30 min) performs the following actions:

(1) *Sensor activation and sampling.* The gas sensors present on the gas board of the node are powered one at a time. After a warm-up period, the first sensor is read, providing a measurement for its correspondent gas. Then, it is switched off and another sensor is powered on (so as to avoid power consumptions peaks). The procedure is repeated for every gas sensor on the board, until all the measurements have been performed. This allows keeping the sensors on for the minimum time requested to take a measurement. The minimum warm-up period necessary to obtain a reliable measurement is indicated for each sensor in the corresponding datasheets. For all the sensors used in this work, we used a warm-up time of 30 s. The overall time required to read all the three sensors used, including the relative warm-up times and acquisition times, is much less than the considered sampling period (30 min).

(2) *Processing and transmission.* At this point, the obtained data are merged together to form a packet, containing also the univocal identifier of the node and a timestamp. Afterwards, the

communication module is initialized. If the Internet connection is available, the sensor node contacts the *uSense* server and immediately sends the string, via *HTTP (HyperText Transfer Protocol)*. Otherwise, if no connection is available, the string is temporarily stored in the SD card (mounted by each node). The stored data will be transmitted to the server in the future, as soon as the connection is available again. This *opportunistic* approach for data communication allows the system to work correctly even when connectivity is intermittent (e.g., the user's WiFi router may be switched off during certain periods).

(3) *Sleep*. Finally, the node disables the communication module and enters the low-power state. The only component that remains active is the real time clock (RTC), that wakes up the MCU *(MicroController Unit)*—through an interrupt—after a predefined time interval (e.g., 30 min). Then, steps 1–3 are performed again.

## *5.2. Wireless Access Points*

After data have been acquired by sensor nodes, they are sent to the server through a wireless access point connected to the Internet. In principle, any wireless technology (e.g., WiFi, ZigBee, Bluetooth, GPRS) can be used. However, we only considered WiFi connectivity in our current implementation. Hence, the user is assumed to have a WiFi access point (or router) covering the area where the sensor node is located.

## *5.3. Server*

The server part of the system consists of three different submodules that accomplish different tasks.

*Database*. It stores information about users, sensor nodes, and measurements taken by sensors. Specifically, for each sensor node, the database contains its geographical coordinates (see below for details) and all the data regarding measured AQI values and gas concentrations.

*Web Server*. It performs two different functions, *i.e.*, *data storage* and *data presentation* to users. First of all, it receives data from sensor nodes and stores them into the database. Specifically, whenever a sensor node has performed a measurement, the acquired data are sent to a dynamic page hosted by the Web server, using the GET method of the HTTP protocol. This page calculates the Air Quality Index and stores all the data in the database together with the identifier of the sensor node and the timestamp.

In addition, the Web server hosts the *uSense* website that allows users to access stored data and view the air quality levels. Specifically, through the Web interface, a user can

(a) create a personal account;

(b) associate sensor nodes to her/his account and set/modify their geographical coordinates (by simply locating them on a map);

(c) select a sensor node (even not belonging to the user) in a map and view the measurements performed in the current day or in a previous one;

(d) view a pollution map, where colored circles indicate the AQI level measured by each sensor (an example is provided in Figure 3);

(e) search for the less polluted path to reach a destination in the city.

*Application Server.* Data stored in the system are also accessible to third-party applications that can elaborate this information to provide additional features and/or statistics. To this end, *uSense* exports a number of Web services, following a *SOA (Service Oriented Architecture)* approach. The exported services are briefly described in Table 3.

**Table 3.** Exported services.

| Service Name | Description |
|---|---|
| *getSensorDataService* | Given a node ID, returns the information contained in the database about that node |
| *getDatafromPeriodService* | Given a node ID and a range of dates, returns all the measurements performed by that node in the specified days |
| *getSensorRangeService* | Given a point (expressed as a pair of coordinates latitude-longitude) and a radius, returns the set of nodes placed in that area |
| *getPollutionDataService* | Given a point and a radius, returns the last measurements performed by nodes in that area |

## 5.4. End-User Devices

Data stored in the system can be accessed by users in several ways. For instance, they can use the Web interface or Web services through a browser or third-party applications, respectively. Furthermore, we have implemented an *app* for smartphones, so as to allow users to quickly and easily access *uSense* even from their mobiles. In this way, users can access the system wherever a data connection is available. The mobile app supplies all the features provided by the Web interface. In addition—exploiting the GPS receiver typically available on smartphones—it is possible to view in a map the sensor nodes closest to the user and check the last measured AQI.

## 6. Implementation

In this section, we provide some implementation details about *uSense*.

## 6.1. Sensor Nodes

The sensor nodes used in *uSense* are *Libelium Waspmote* (Figure 2a). They are equipped with an 8-bit microcontroller, and a WiFi module that allows TCP/IP and UDP/IP socket connections and HTTP/HTTPS communications. Each sensor node is provided with a gas sensor board, where CO, $NO_2$, $O_3$, temperature and humidity sensors have been mounted (specifically, *TGS2442*, *MiCS-2714* and *MiCS-2614* sensors for CO, $NO_2$ and $O_3$, respectively). In our prototype system, we have considered a sampling period of 30 min (*i.e.*, two samples per hour) for each gas under test.

Waspmote nodes do not run any operating system. They can be programmed in *C* through a dedicated *IDE (Integrated Development Environment)*, exploiting some libraries provided by the producer to interface with sensors, microcontroller and WiFi module. To allow easily deploying sensor nodes outside, we have realized a PVC box for each sensor node (shown in Figure 2b), containing—in addition to the Waspmote—an antenna, an activation button, a 6600 mAh rechargeable battery and a small brushless fan powered by an additional battery (so as to allow the air circulation). The use of a PVC housing may cause artifacts since this material can degrade in high concentrations of $O_3$ and $NO_2$, so that interaction between the gas molecules and the housing surface may be supposed. Indeed, this choice was motivated to minimize the cost of the single nodes, making future deployment of very dense monitoring systems economically sustainable. Use of highly inert plastics, such as *Polytetrafluoroethylene (PTFE)*, which would considerably increase the cost of individual nodes, will be considered for comparison purposes in future experiments.

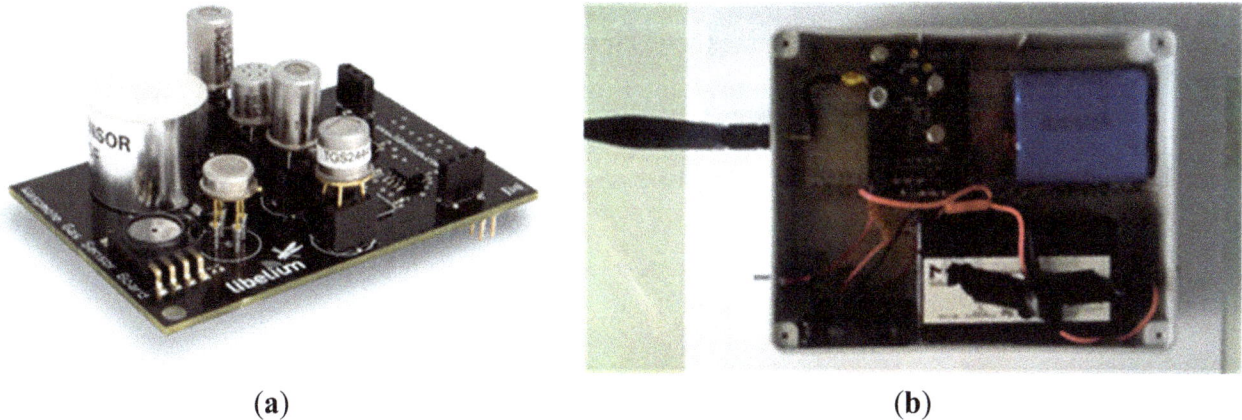

(a)                                                  (b)

**Figure 2.** (**a**) Libelium Gas Sensor Board. (**b**) A sensor node inside its PVC box.

In order to obtain an acceptable accuracy, each sensor node should be individually calibrated. The calibration phase is often overlooked in articles regarding sensor networks. Unfortunately, a common characteristic of the miniaturized and inexpensive sensors that are typically mounted on sensor nodes is the large parameter tolerance. This means that the response curve of different samples of the same kind of sensor (e.g., $NO_2$ sensor) prior calibration are so poorly matched that comparison of data produced by distinct sensor nodes are generally meaningless. For this reason, we preliminarily performed two-point calibration of all the sensors used in this work. Note that all the considered sensors are of resistive type and their responses are nearly linear in a log-log plot, as reported in datasheets. For this reason, the following expression was considered for the calculation of the gas concentration:

$$log_{10}(C) = k_1 log_{10}(R) + k_2 \tag{3}$$

where $C$ is the gas concentration, $R$ the sensor resistance, whereas $k_1$ and $k_2$ are constants depending on the sensor type. The calibration procedure consisted in exposing the whole sensor nodes to different gas concentrations inside a small sealed chamber. The first step consisted in determining the optimal value of load resistance, which is a resistor placed in a series with the sensor resistance in order to form a voltage divider. The input voltage for the divider is a constant reference voltage, so that the output voltage is a function of the sensor resistance. An initial value for the load resistance was selected using the information provided by the sensor datasheets; this value is used to estimate the sensor resistance in the reference concentration. The final value of the load resistance is chosen in order to maximize the estimated output voltage swing across the operating concentration range. This is an important step, considering the relatively low-resolution AD converters (10 bit) that equip the sensor nodes. Tests were practically performed using two gas concentrations: (i) zero pollutant level, obtained exposing the sensor to synthetic air and (ii) reference level (*i.e.*, 1.45 ppm for $NO_2$ and 100 ppm for CO) obtained by using gas cylinders, containing a known concentration of the pollutant in synthetic air. After setting a proper load resistance through direct programming of the sensor firmware, measurement of sensor resistance in the two concentration conditions allows to determine the constants $k_1$ and $k_2$ in Equation (3). Since a zero concentration is not representable with Equation (3), we have assumed that concentration in synthetic air is the one for which the sensor resistance reaches the value in pure air (reported in the datasheets). All measurements were taken after a delay of 30 s after switching on the sensors: this was necessary to allow the sensors to reach their operating temperature, as described in the sensor documentation. A gas mixture with precise $O_3$ concentration was not available so that calibration of the $O_3$ sensor was performed using only the synthetic air measurement (single-point calibration).

The installation of a sensor node is quite straightforward. The user has only to digit the passphrase of her/his WiFi network inside a specific configuration file in the SD of the sensor node. Then, she/he has to create a personal account on the *uSense* website and associate the sensor node with the newly created account, by indicating where it is placed on a map.

*6.2. Server*

The server part of the system is implemented as three distinct software processes (see Section 5.3). In particular, we used *PostgreSQL* as object-relational database management system. In addition, we resorted to *PostGIS*, a specific spatial extension of PostgreSQL, to handle geographical objects and queries (*i.e.*, involving latitude and longitude coordinates).

The used Web server is *Apache* and the dynamic pages have been written in *PHP*. The website uses the *Google Maps API* to visualize sensor nodes and AQI levels on maps and determine the shortest paths between two points. Plots and graphs have been realized through the *Chart.js* JavaScript library. A screenshot of the website is shown in Figure 3.

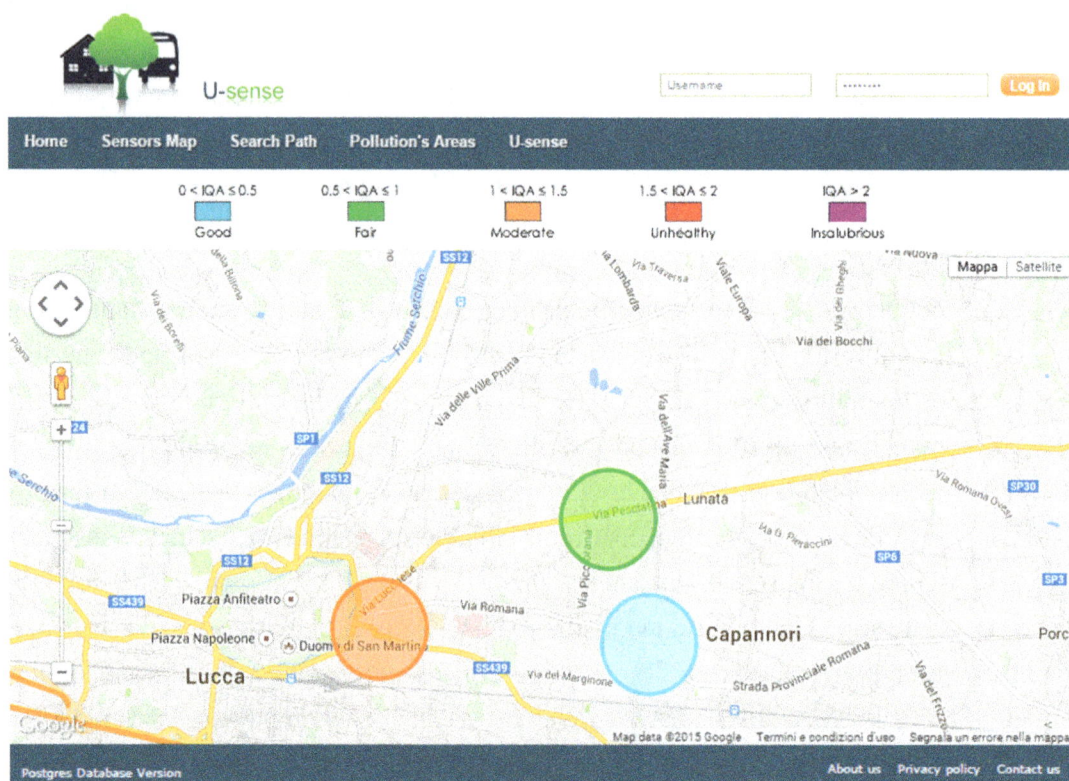

**Figure 1.** Webpage showing AQI levels for three sensor nodes.

Finally, *JBoss* is the application server used to export Web services. Using JBoss—and, thus, *Java EE (Enterprise Edition)*—assures performance, scalability, security, and reliability. The web services are called through the *SOAP* protocol, over HTTP. The results of the implemented services are returned as *JSON* strings.

## 6.3. End-User Devices

The smartphone app has been realized for *Android*. Figure 4-left shows the initial page of the *uSense* app and the available choices. The screenshot in Figure 4-right shows the sensor nodes located near to the user. To implement this feature, we used the *Google Maps API* and the *GPS* receiver integrated in the smartphone.

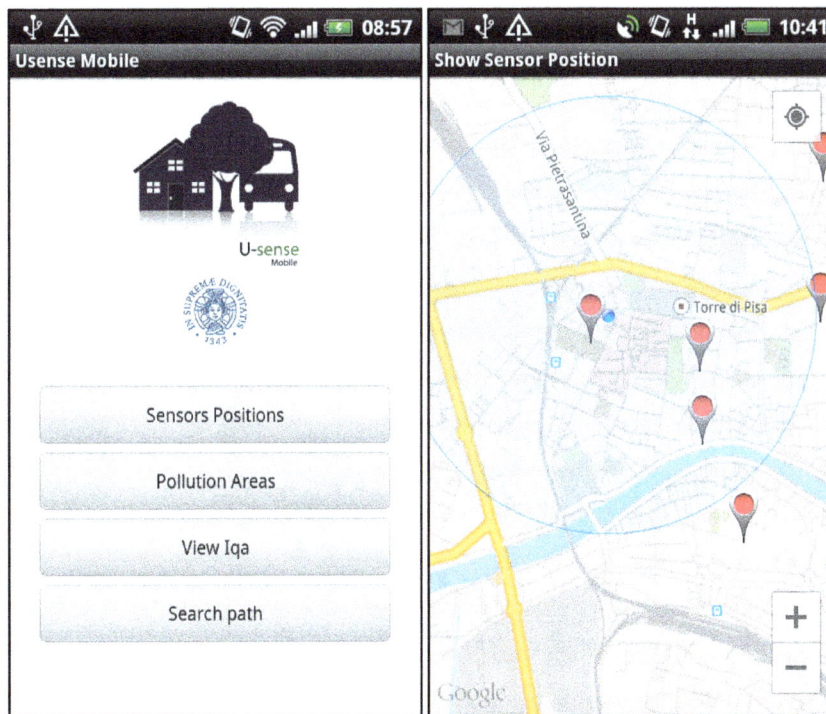

**Figure 4.** Mobile interface: initial page (**Left**) and nearest sensor nodes (**Right**).

## 7. Experimental Results

In order to test the overall system, we performed an in-field experimentation of *uSense*. To this end, we installed a number of sensor nodes in different urban areas. In the following, we will refer to the experimental measurements obtained from three sensor nodes installed in the area of Lucca—a city in Tuscany—in strategic location with different expected pollution levels. Sensor nodes—inside their PVC boxes—were installed near roads, in order to investigate the impact of traffic on air quality. Specifically, we considered three areas (namely *Zone A*, *Zone B*, *Zone C*) with different traffic conditions and, hence, different expected pollution levels. Zone A is near the center of the city and is characterized by heavy traffic. On the contrary, Zone B is in the countryside and the sensor node is located near a rural road. Finally, Zone C is a suburban area with medium traffic intensity. Figure 5 shows the positions on the map of the three sensor nodes during the in-field experimentation, whereas Figure 6 shows the sensor nodes inside their boxes, as they have been placed in the various zones.

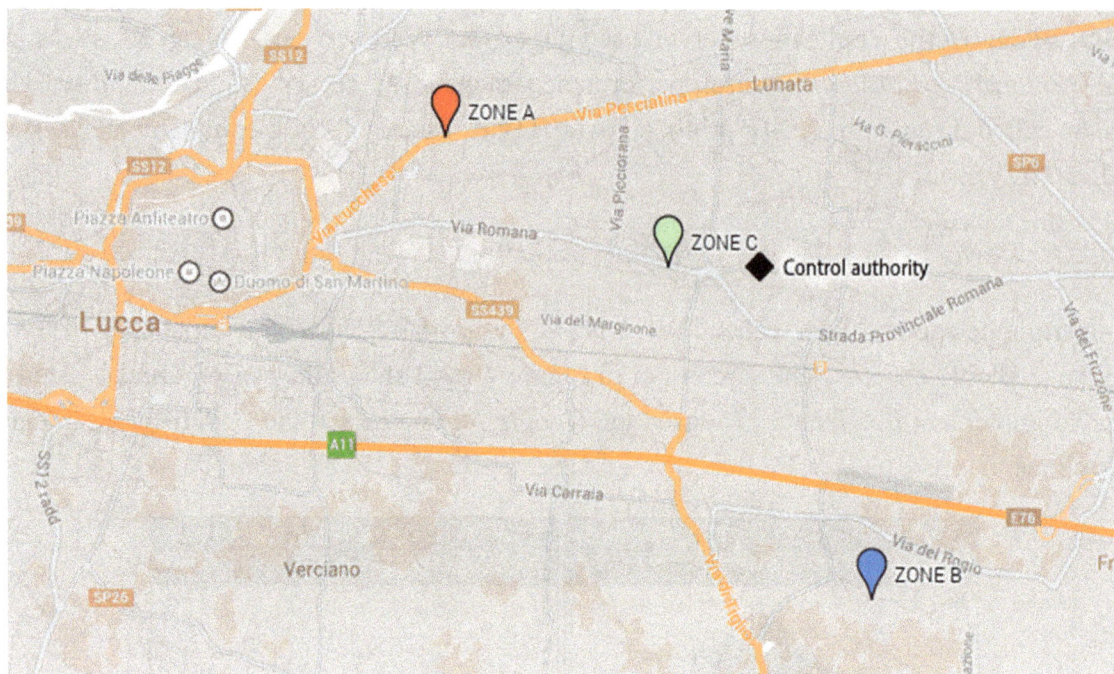

**Figure 5.** Locations of sensor nodes and control authority sensing station during in-field experimentation.

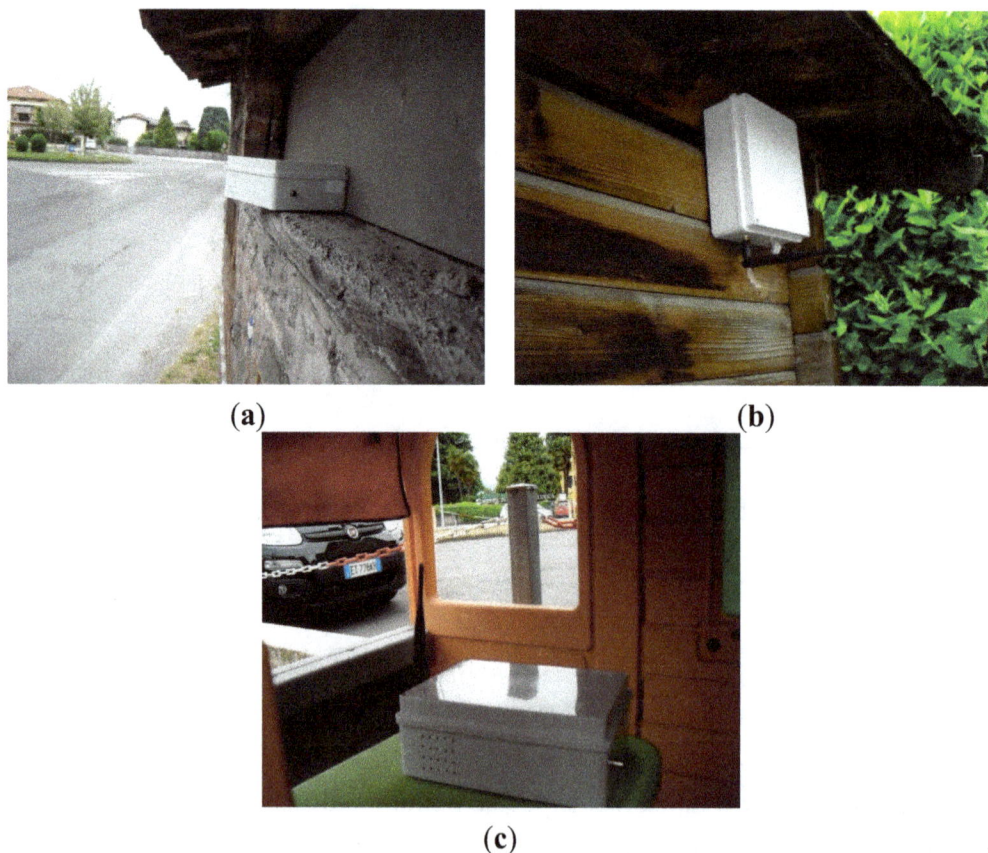

(a)                                                                    (b)

(c)

**Figure 6.** Sensor installation in Zone A (**a**), Zone B (**b**), Zone B (**c**).

The monitoring phase started on 1 May 2014 and ended on 1 June 2014. Since nodes were placed outdoors, the WiFi signal strength was quite variable, and, therefore, the WiFi connection was not optimal. However, we observed that no measurement was missed, thanks to the reliable data transfer service implemented in *uSense*. The experimental results are plotted in Figures 7 and 8.

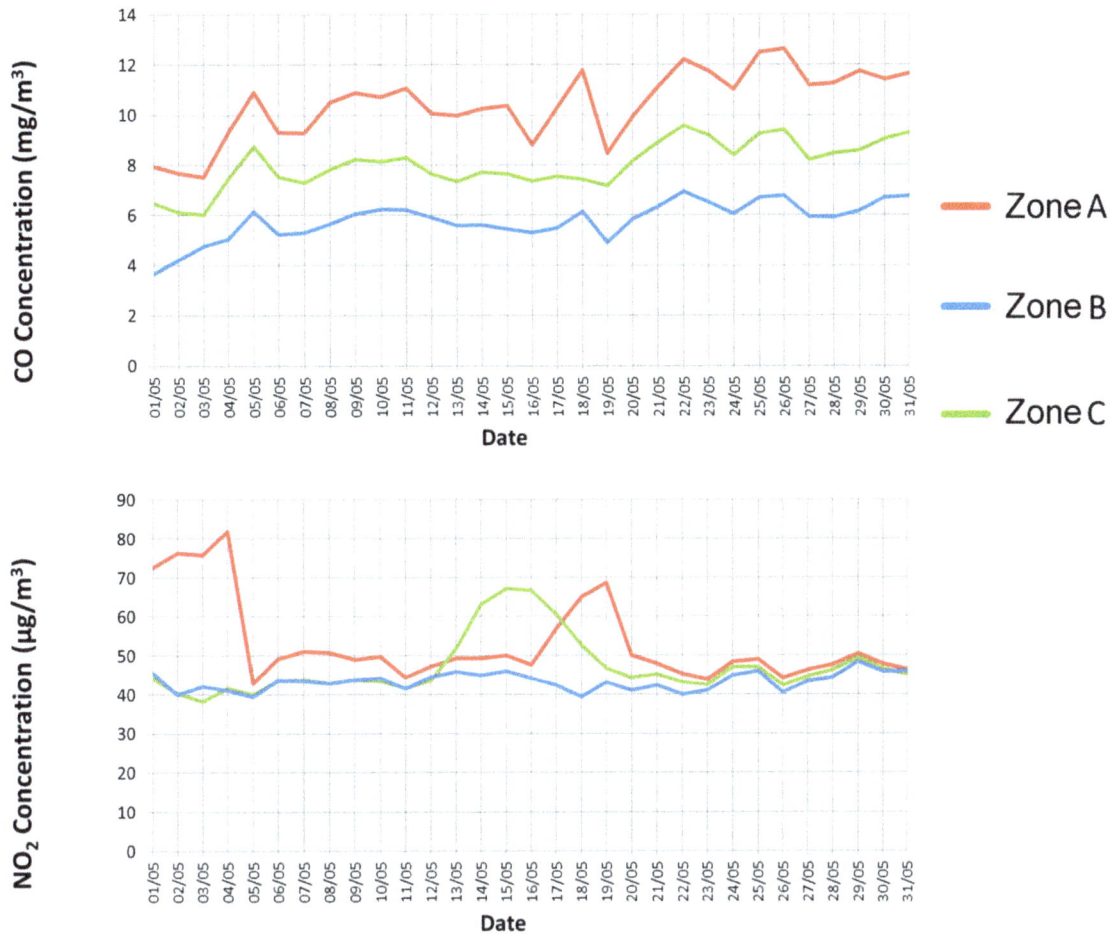

**Figure 7.** Experimental results: gas concentrations (NO$_2$, and CO) in the three zones.

Figure 7 shows, for each considered day, the maximum average concentration of CO, and NO$_2$. In fact, these gases are the most significant ones to monitor the pollution due to traffic, since they are mainly produced, through combustion, by engines that run on fossil fuel, such as petroleum derivatives. Specifically, we show the daily maximum 8-h average value for CO and the maximum hourly average value for NO$_2$ (similarly to how reference values are expressed in Italian regulations). Instead, Figure 8 reports the measured AQI values, according to the definition given in Section 4. The obtained results confirm our expectations. Zone A is the most polluted area, due to the transit of many vehicles, in particular trucks, and the lack of vegetation. Conversely, Zone B—a rural and low-traffic area—exhibits low pollution levels, which are absolutely negligible. Zone C, instead, shows intermediate gas concentrations, but which are still acceptable.

To validate the collected results, we have compared them with the values measured by the local environmental control authority (namely ARPAT) for the entire period. All the data gathered by the environmental control authority are daily published and freely available in the ARPAT website [16]. Among the various monitoring stations of the control authority placed near the considered city, we have chosen the closest to the *uSense* sensor nodes. Specifically, we have compared the measurements obtained by the sensor node in Zone C with the ones provided by the ARPAT sensing station named LU-CAPANNORI (whose location is shown in Figure 5). The node in Zone C is distant less than 800 m from the sensing station of the control authority. Since the latter does not sample CO and O$_3$, we have limited our comparison to $NO_2$. As clearly emerges from Figure 9, our results match closely the values provided by the environmental control authority during the whole duration of the experiment.

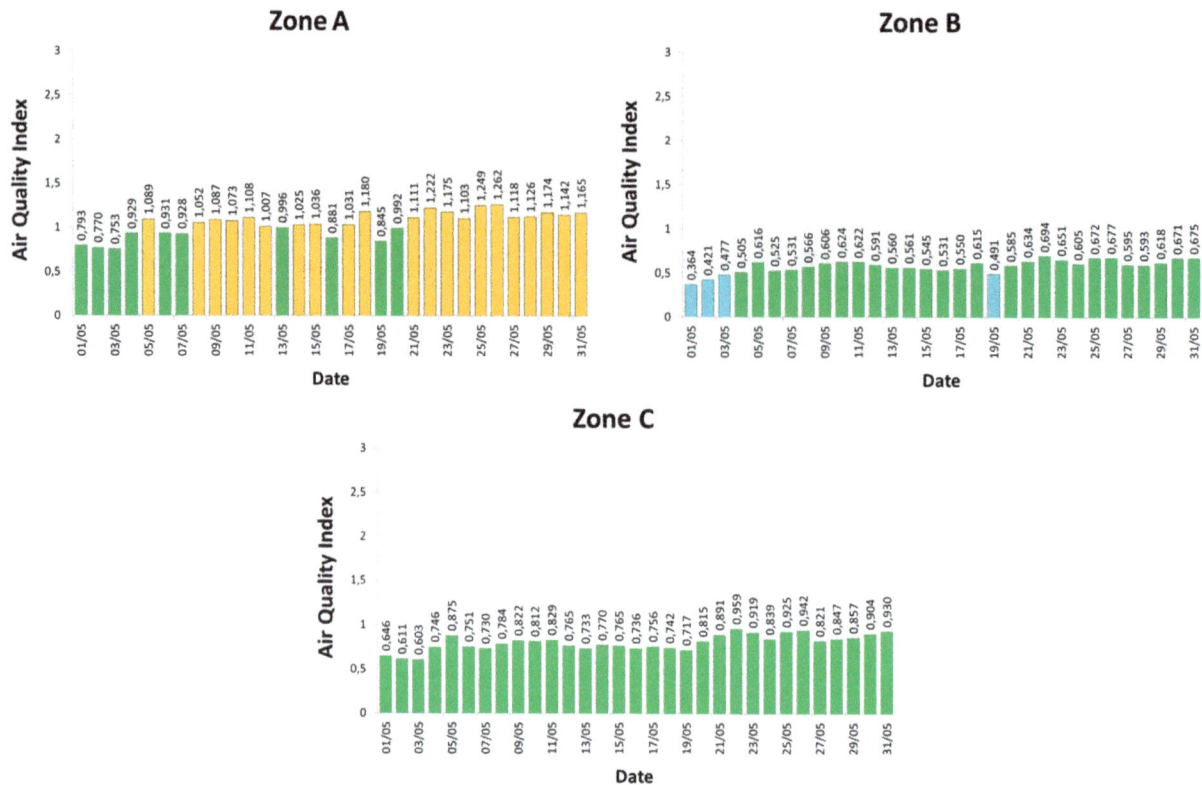

**Figure 8.** Experimental results: Air Quality Index (AQI) in the three zones.

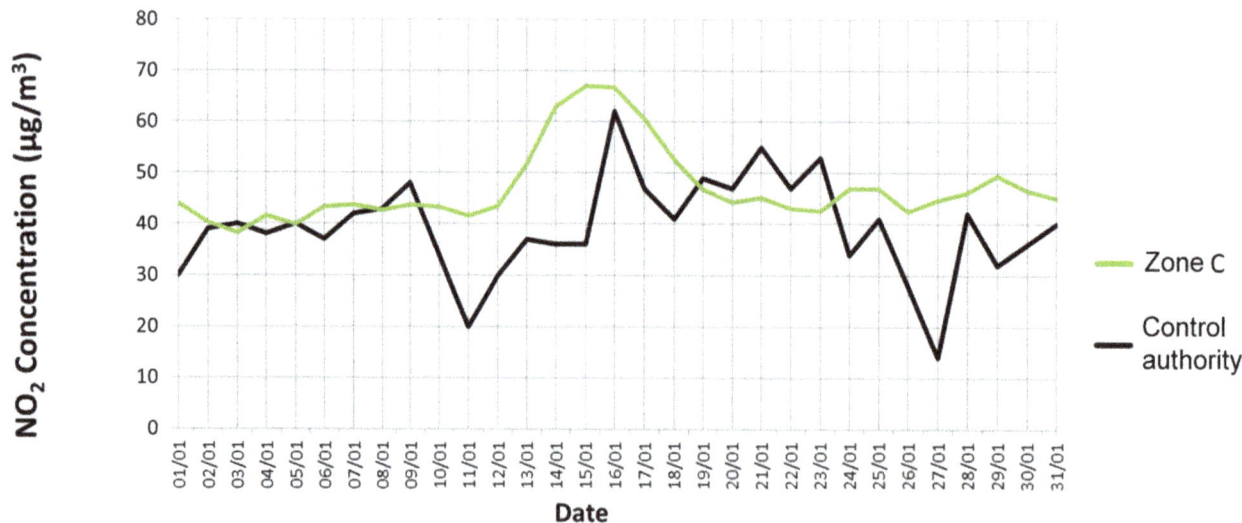

**Figure 9.** Experimental results: $NO_2$ values measured in Zone C and values provided by the local environmental control authority.

## 8. Conclusions

In this paper, we have presented *uSense*, a low-cost sensing system for cooperative air quality monitoring in urban areas. The novel contribution of *uSense* is represented by the possibility for users to monitor the air pollution level inside their properties, by simply installing one or more sensor nodes. *uSense* follows a cooperative sensing approach, in the sense that the data collected by each sensor node are uploaded to a remote server and shared among the users of the system, so that each one can obtain a more complete and reliable information about the pollution of the various parts of her/his city. Data collected by *uSense* can be accessed in several ways, *i.e.*, through a Web Interface, a Mobile App or Web Services.

Compared to other solutions present in the literature, *uSense* relies on small, low-power sensor nodes, provided with long-lasting rechargeable batteries (6600 mAh) and WiFi communication modules. This way, they can be easily moved and deployed outside, wherever there is WiFi coverage. For instance, they can be placed in a balcony, in a garden, on a window sill or can be hung to a wall. A sensor node can work even when the network connectivity is intermittent, so as to avoid data loss. In this case, in fact, the node can temporarily store the collected data in a SD card, waiting for the connection to be available again. In addition, given the modularity of the used *Libelium* board, extending the range of the monitored pollutant is straightforward.

*uSense* has been tested through an in-field experiment, in three different urban areas. The experimental measurements—compared with the values provided by the local environmental control agency—show that, even though measurements obtained from low-cost sensors are not as accurate as official data, they can still provide useful indications of air quality in a specific location.

Also, it may be worthwhile emphasizing that the data quality provided by *uSense* depends on the accuracy of low-cost sensors. As the technology evolves, more accurate low-cost sensors will be made available and the data quality will improve accordingly. From this perspective, *uSense* can be considered an evolving low-cost monitoring platform whose data quality can benefit from developments in sensor technology.

## Acknowledgments

This work has been partially supported by the Tuscany Region, in the framework of the SMARTY project (POR program 2007-13), and partially by the University of Pisa, in the framework of the PRA 2015 program. Special thanks to Maria Gabriella Antona who implemented some components of the system.

## Author Contributions

Simone Brienza has designed the system architecture and developed the system. Andrea Galli has programmed and deployed the sensor nodes, conducted the experiments, and collected the data. Giuseppe Anastasi has supervised the whole work. Paolo Bruschi has also supervised the work with special focus on sensors and sensor calibration.

## Conflicts of Interest

The authors declare no conflict of interest.

## References

1.  European Environment Agency. European Union emission inventory report 1990–2012 under the UNECE Convention on Long-range Transboundary Air Pollution (LRTAP). June 2014. Available online: http://www.eea.europa.eu/publications/lrtap-2014 (accessed on 21 January 2015).

2.  United States Environmental Protection Agency. 1970–2013 Average annual emissions, all criteria pollutants in MS Excel., February 2014. Available online: http://www.epa.gov/ttn/chief/trends/trends06/national_tier1_caps.xlsx (accessed on 21 January 2015).

3.  European Environment Agency. Air quality in Europe—2014 Report. November 2014. Available online: http://www.eea.europa.eu/publications/air-quality-in-europe-2014 (accessed on 21 January 2015).

4.  Kampa, M.; Castanas, E. Human health effects of air pollution. *Environ. Pollut.* **2008**, *151*, 362–367.

5.  Rana, R.; Chou, C.T.; Bulusu, N.; Kanhere, S.; Hu, W. Ear-Phone: A context-aware noise mapping using smart phones. *Pervasive Mobile Comput.* **2015**, *17*, 1–22.

6.  Kim, S.; Paulos, E. InAir: Sharing indoor air quality measurements and visualizations. In Proceedings of the ACM SIGCHI Conference on Human Factors in Computing Systems, Atlanta, GA, USA, 10–15 April 2010.

7.  Jiang, Y.; Li, K.; Tian, L.; Piedrahita, R.; Yun, X.; Mansata, O.; Lv, Q.; Dick, R.P.; Hannigan, M.; Shang, L. MAQS: A personalized mobile sensing system for indoor air quality monitoring. In Proceedings of the 13th ACM International Conference on Ubiquitous Computing, Beijing, China, 17–21 September 2011.

8.  Nikzad, N.; Verma, N.; Ziftci, C.; Bales, E.; Quick, N.; Zappi, P.; Patrick, K.; Dasgupta, S.; Krueger, I.; Rosing, T.S.; *et al.* CitiSense: Improving geospatial environmental assessment of air quality using a wireless personal exposure monitoring system. In Proceedings of the ACM Conference on Wireless Health, San Diego, CA, USA, 23–25 October 2012.

9.  Dutta, P.; Aoki, P.M.; Kumar, N.; Mainwaring, A.; Myers, C.; Willett, W.; Woodruff, A. Common Sense: Participatory urban sensing using a network of handheld air quality monitors. In Proceedings of the 7th ACM Conference on Embedded Networked Sensor Systems, Berkeley, CA, USA, 4–6 November 2009.

10. Hasenfratz, D.; Saukh, O.; Sturzenegger, S.; Thiele, L. Participatory air pollution monitoring using smartphones. In Proceedings of 2nd International Workshop on Mobile Sensing, Beijing, China, 16 April 2012.

11. Hasenfratz, D,; Saukh, O.; Walser, C.; Hueglin, C.; Fierz, M.; Arn, T.; Beutel, J.; Thiele, L. Deriving high-resolution urban air pollution maps using mobile sensor nodes. *Pervasive Mobile Comput.* **2015**, *16*, 268–285.

12. Brienza, S.; Galli, A.; Anastasi, G.; Bruschi, P. A cooperative sensing system for air quality monitoring in urban areas. In Proceedings of 2014 International Conference on Smart Computing Workshops (SMARTCOMP Workshops), Hong Kong, 5 November 2014.

13. U.S. Environmental Protection Agency (EPA). Air Quality Index—A Guide to Air Quality and Your Health, February 2014. Available online: http://www.epa.gov/airnow/aqi_brochure_02_14.pdf (accessed on 30 April 2015).

14. CITEAIR II (Common Information to European Air). Indices definition. Available online: http://www.airqualitynow.eu/about_indices_definition.php (accessed on 30 April 2015).

15. Directive 2008/50/EC of the European Parliament and of the Council on ambient air quality and cleaner air for Europe, 21 May 2008. Available online: http://eur-lex.europa.eu/legal-content/EN/TXT/?uri=CELEX:32008L0050 (accessed on 06 May 2015).

16. ARPAT, Agenzia Regionale per la Protezione Ambientale della Toscana (Regional Agency for Environmental Protection of Tuscany). Bollettino regionale di qualità dell'aria (Regional air quality bulletin). Available online: http://www.arpat.toscana.it/apps/bollaria/f?p=bollaria:regionale: (accessed on 5 May 2015).

# 4

# Smartphone Applications with Sensors Used in a Tertiary Hospital—Current Status and Future Challenges

**Yu Rang Park [1,†], Yura Lee [2,†], Guna Lee [3], Jae Ho Lee [2,4,5,6] and Soo-Yong Shin [2,4,*]**

[1]  Clinical Research Center, Asan Medical Center, Seoul 138-736, Korea;
E-Mail: yurang.park@amc.seoul.kr

[2]  Department of Biomedical Informatics, Asan Medical Center, Seoul 138-736, Korea;
E-Mails: haepary@naver.com (Y.L.); rufiji@gmail.com (J.H.L.)

[3]  Division of Nursing Science, College of Health Science, Ewha Womans University, Seoul 120-750, Korea; E-Mail: finside00@gmail.com

[4]  Ubiquitous Health Center, Asan Medical Center, Seoul 138-736, Korea

[5]  Department of Emergency Medicine, Asan Medical Center, University of Ulsan College of Medicine, Seoul 138-736, Korea

[6]  Division of General Internal Medicine and Primary Care, Brigham and Women's Hospital, Boston, MA 02467, USA

[†]  These authors contributed equally to this work.

[*]  Author to whom correspondence should be addressed; E-Mail: sooyong.shin@amc.seoul.kr

Academic Editor: Ki H. Chon

**Abstract:** Smartphones have been widely used recently to monitor heart rate and activity, since they have the necessary processing power, non-invasive and cost-effective sensors, and wireless communication capabilities. Consequently, healthcare applications (apps) using smartphone-based sensors have been highlighted for non-invasive physiological monitoring. In addition, several healthcare apps have received FDA clearance. However, in spite of their potential, healthcare apps with smartphone-based sensors are mostly used outside of hospitals and have not been widely adopted for patient care in hospitals until recently. In this paper, we describe the experience of using smartphone apps with sensors in a large medical center in Korea. Among >20 apps developed in our medical center, four were extensively analyzed ("My Cancer Diary", "Point-of-Care HIV Check", "Blood

Culture" and "mAMIS"), since they use smartphone-based sensors such as the camera and barcode reader to enter data into the electronic health record system. By analyzing the usage patterns of these apps for data entry with sensors, the current limitations of smartphone-based sensors in a clinical setting, hurdles against adoption in the medical center, benefits of smartphone-based sensors and potential future research directions could be evaluated.

**Keywords:** healthcare app; data entry; mobile health; smartphone sensors

## 1. Introduction

The number of global smartphone subscribers is expected to reach 3.5 billion by 2019 [1]. Due to their popularity, processing power, non-invasive and cost-effective sensors, and wireless communication capabilities, smartphones have received great attention in healthcare settings [2–5]. A smartphone was used by 74% of physicians for professional purposes in 2013 [4]. As a consequence, the mobile health (m-health) market has grown rapidly and will reach $49 billion by 2020 [6]. Patients and consumers also expect that m-health will improve the quality and convenience of healthcare by reducing costs [7]. Hence, many healthcare applications (apps) have now been developed and commercialized (*i.e.*, ECG monitoring and atrial fibrillation detection [8], eye disease detection [9]). Several apps have also received FDA clearance [10]. Additionally, in the United States, the Medical Electronic Data Technology Enhancement for Consumers' Health (MEDTECH) Act was recently proposed to exempt low-risk medical software and mobile apps from FDA regulation [11].

Among the enormous number of healthcare apps available for smartphones, some apps use diverse sensors for data entry. The sensors for smartphones can be classified into two categories including smartphone-based sensors and add-on sensors. Smartphone-based sensors include the cameras, accelerometers, gyroscopes, magnetometers, proximity sensors, light sensors, barometers, thermometers, air humidity sensors, and pedometers which are integrated into most smartphones [12]. Some recent smartphones have additional biometric sensors including heart rate monitors and fingerprint sensors [12]. Add-on sensors use external devices such as smartphone cases [8]. Most apps with sensors have been highlighted for non-invasive physiological monitoring outside of hospitals. In spite of their promise, healthcare apps with sensors have not been widely adopted for patient care within hospitals until recently [3,13]. Most apps are designed for reference or education purposes only [13].

In our current study, we report our experiences of using smartphone apps with smartphone-based sensors in our hospital, the Asan Medical Center (AMC). AMC is the largest medical center in Korea with approximately 2700 inpatient beds and 10,000 outpatient visits per day. Since its establishment, the hospital information system has been actively used to improve quality of care and to make the clinical workflow more efficient [14]. In the early 2000s, the AMC began using m-health services such as a mobile electronic medical record (EMR) system to improve the accessibility, mobility, and efficiency of patient care. The AMC also established a ubiquitous health center in 2009 to promote its m-health service in the hospital [15]. Due to the very large number of patients and the active use of the m-health service, our cases may be helpful to other hospitals trying to adopt or expand m-health.

## 2. Materials and Methods

### 2.1. Selection of AMC Healthcare Apps

Among more than 20 healthcare apps which have been developed at the AMC [15], four apps—"My Cancer Diary", "Point-of-Care HIV Check", "Blood Culture" and "mAMIS (mobile Asan Medical Information System)"—were chosen due to their use of smartphone-based sensors for data entry. The remaining apps do not have the functionality to enter the data from a sensor, but only to provide the necessary information to users. Table 1 summarizes the characteristics of these four apps. "My Cancer Diary" was developed for patient use and other apps were designed for use by healthcare providers.

**Table 1.** Characteristics of the chosen apps.

|  | My Cancer Diary | Point-of-Care HIV Check | Blood Culture | mAMIS |
|---|---|---|---|---|
| Description | Personal health management application for cancer patients | Point-of-Care HIV check | Point-of-Care blood culture sampling application | Mobile electronic medical record system |
| Target users | Patients | Nurses | Physicians | Physicians and nurses |
| Period of usage | Since October 2012 | Since September 2013 | Since May 2012 | Since September 2014 * |
| Type of sensor | Camera (barcode or QR code recognition) | Camera (barcode recognition and image acquisition) | Camera (barcode recognition) | Touch ID (Fingerprint recognition) |
| Purpose of sensor | Log-in | Patient identification, result image acquisition | Patient and blood sample identification | Log-in |
| Supported OS | Android, iOS | Android | Android, iOS | iOS (8.0 and up) |

\* After supporting Touch ID for iPhone 6 and later.

### 2.1.1. My Cancer Diary

The "My Cancer Diary" app is a tool for cancer patient self-management. Usually, patient self-management tools provide information or functions such as general information about the disease, patient assistance, and healthcare professional assistance tools to aid in patient self-management of their chronic diseases or symptoms [16]. Following these trends, "My Cancer Diary" provides three types of information: (1) general information about cancer (anticancer drugs, serious symptoms, frequently asked questions about cancers, and patients' essays on how they overcame their cancers); (2) a patient assistance tool (symptom management and my anticancer diary); and (3) healthcare professional assistance (cancer education). As mentioned in Table 1, "My Cancer Diary" used the camera to recognize the patient's barcode or QR code for app login. Figure 1 shows the barcode login image (Figure 1A) and main menu of the app (Figure 1B).

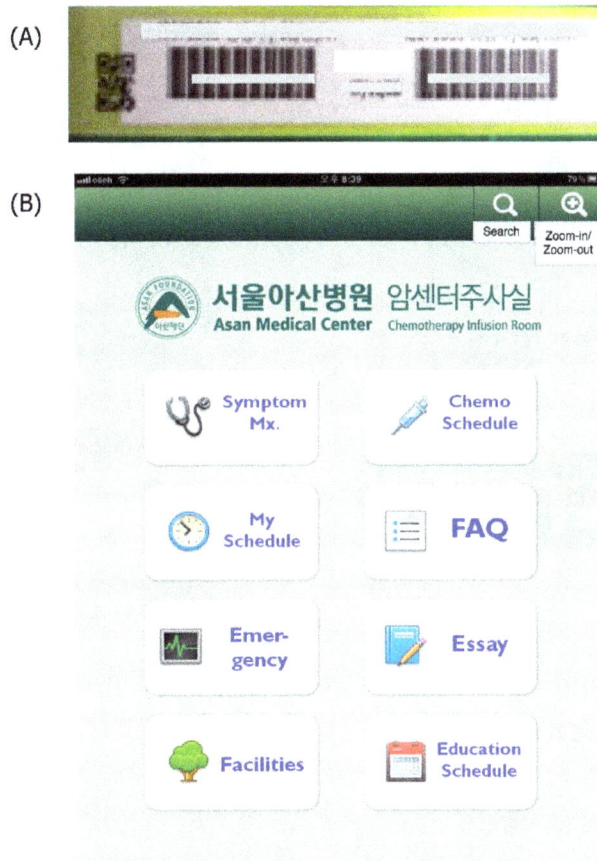

**Figure 1.** My Cancer Diary application **(A)** Patient barcode sample **(B)** Screenshot of main menu.

## 2.1.2. Point-of-Care HIV Check

Point-of-care diagnosis apps are essential to improve patient and care provider safety. A point-of-care diagnosis app in the AMC checks HIV infection of patients in the emergency room (ER) to improve care provider safety. In the ER, physicians and nurses contact many patients without previous laboratory studies to evaluate for possible infectious diseases, and 7% of occupationally acquired HIV infections occur in the ER [17,18]. For example, approximately 100,000 sharps injuries are reported from NHS hospitals in the United Kingdom. In the United States, approximately 400,000 sharps-related injuries among health care personnel are reported annually, and that number is considered an underestimate. Therefore, a fast and easily accessible report of the result of an HIV screening test is required to provide extra care to patients highly likely to be HIV-positive [19].

The developed "Point-of-Care HIV Check" app provides a local, rapid, and often lower-cost alternative to sending samples to a hospital laboratory for analysis [20]. The app identifies a patient's ID using a barcode and takes a photo of the test results using the smartphone camera. Figure 2A shows the main menu of the app, and Figure 2B shows the barcode reading phase.

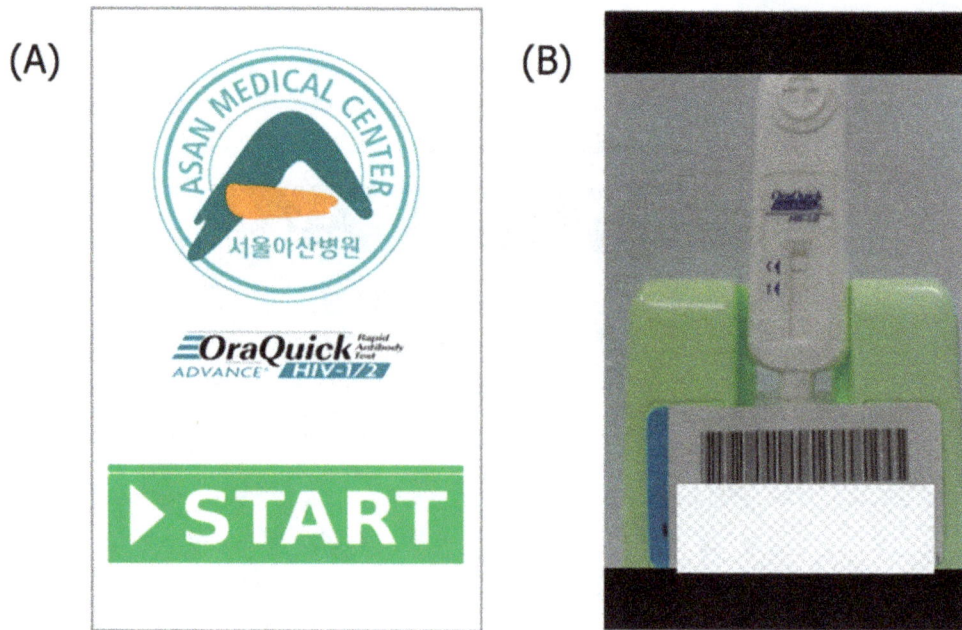

**Figure 2.** "Point-of-care HIV Check" application. (**A**) Screenshot of main menu; (**B**) Screenshot of barcode reading phase.

### 2.1.3. Blood Culture

When physicians sample a patient's blood for blood culture examination, they should sample the blood two or more times with a time gap between measurements [21]. The "Blood Culture" app was implemented to guide and alarm at the proper method and time to sample blood for culture. This app identifies the patient's barcode using the smartphone camera and creates a timestamp for the blood culture [22]. Users must login to this app using a hospital staff ID and scan the patient and sample barcodes at each step (Figure 3A,B).

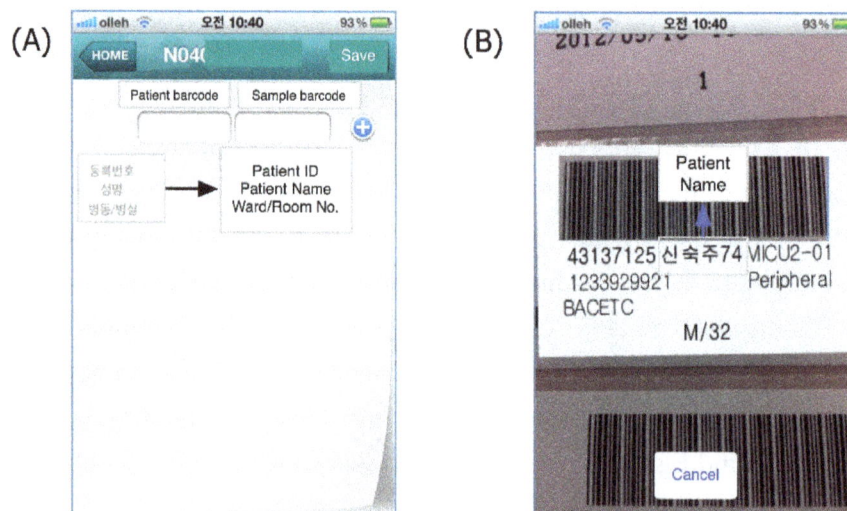

**Figure 3.** *Cont.*

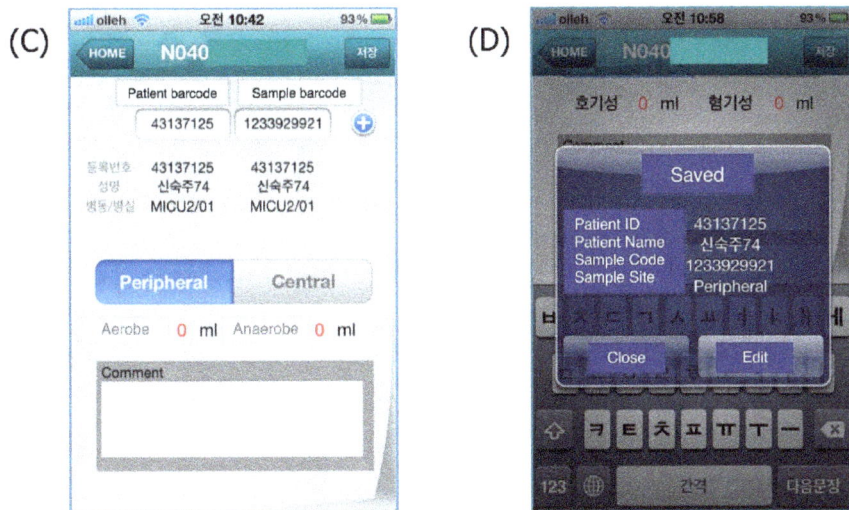

**Figure 3.** "Blood Culture" application. (**A**) Barcode reading menu; (**B**) Barcode reading phase; (**C**) Reading results and entering additional data such as the sample site and volume; (**D**) Saving data. In this figure, pseudonymized ID and patient name were used.

Afterwards, the sample site and volume can be entered manually (Figure 3C). These data are then transmitted to the legacy laboratory information system simultaneously (Figure 3D).

2.1.4. mAMIS

mAMIS is a mobile EMR app which provides all medical records including medications, laboratory results, and images of radiologic studies completed since 2004 (Figure 4).

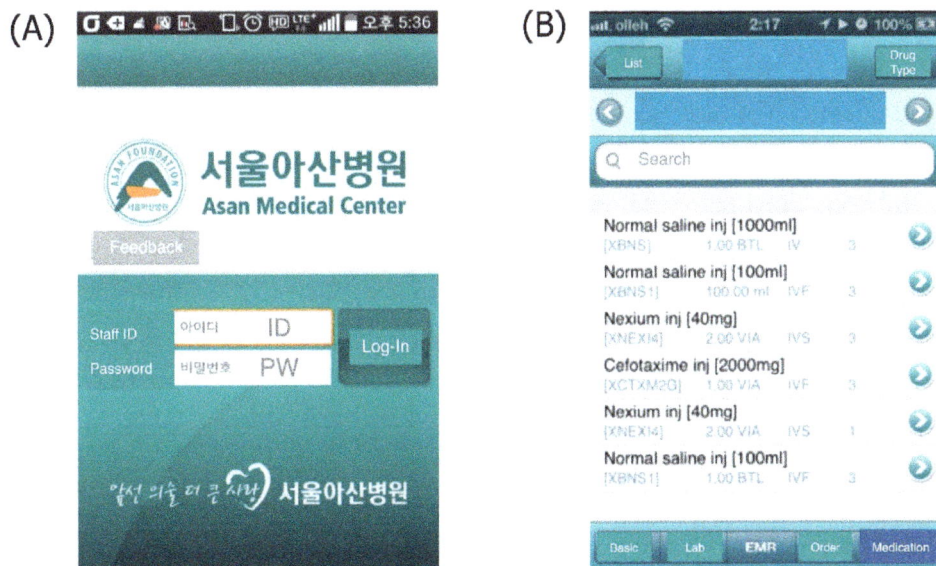

**Figure 4.** *Cont.*

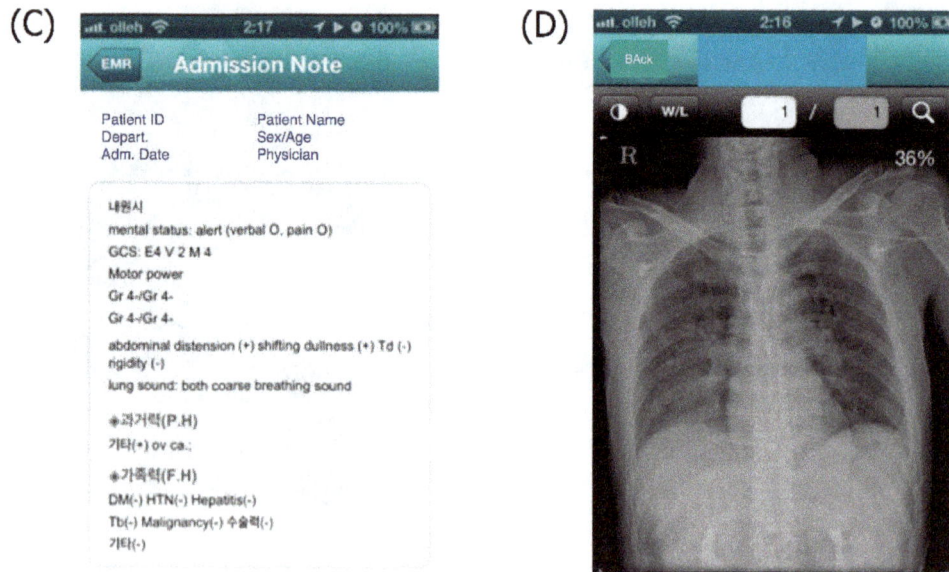

**Figure 4.** mAMIS application. (**A**) mAMIS login menu; (**B**) Medication information for the chosen patient; (**C**) Admission note; (**D**) PACS image.

The app has been widely used since 2010, and a sensor for fingerprint recognition was added in September 2014. After introduction of Touch ID into Apple iPhone 6, the AMC developed the login capability of the mAMIS iOS version due to the large number of requests received.

## 2.2. Collection and Analysis of Usage Data

To check the current status of mobile sensor usage of healthcare apps in the AMC, the log data of each app were collected. Since all of the chosen apps were developed in-house, the necessary log data was easily extracted from the mobile application server or hospital information system server. The log data from the apps' launch until November 2014 were collected for "My Cancer Diary", "Point-of-Care HIV Check" and "Blood Culture" apps. The log data of "My Cancer Diary" was collected for 26 months, "Point-of-Care HIV Check" for 14 months, and "Blood Culture" for 29 months. The log data for "mAMIS" was collected from the app upgrade to support Touch ID (September 2014) until March 2015. Since the other apps have more than one year log data, we expanded the log data collection period for mAMIS (seven months). To represent the trend of mobile sensor usage of the four apps, we analyzed the log data using linear regressions and added the results as trend lines in the figures.

## 2.3. System Architecture of AMC Mobile Applications

The four selected apps have the same technical architecture (Figure 5). We implemented a broker server for mobile apps to enforce the security of clinical data in legacy system. This broker server prohibits direct access to the legacy clinical database by mobile client application. It provides Secure Socket Layer (SSL) and data encryption function, and Jolt-based secure communication channel to the legacy database [23]. Finally, the captured clinical data stored in the legacy database pass through the data control layer. The mobile client application comprises of following five layers: Data input layer,

Presentation layer, Business layer, Data layer, and Network layer. Clinical data from sensors on smartphone captured and handled by above five layers. Then captured clinical data were sending to broker server.

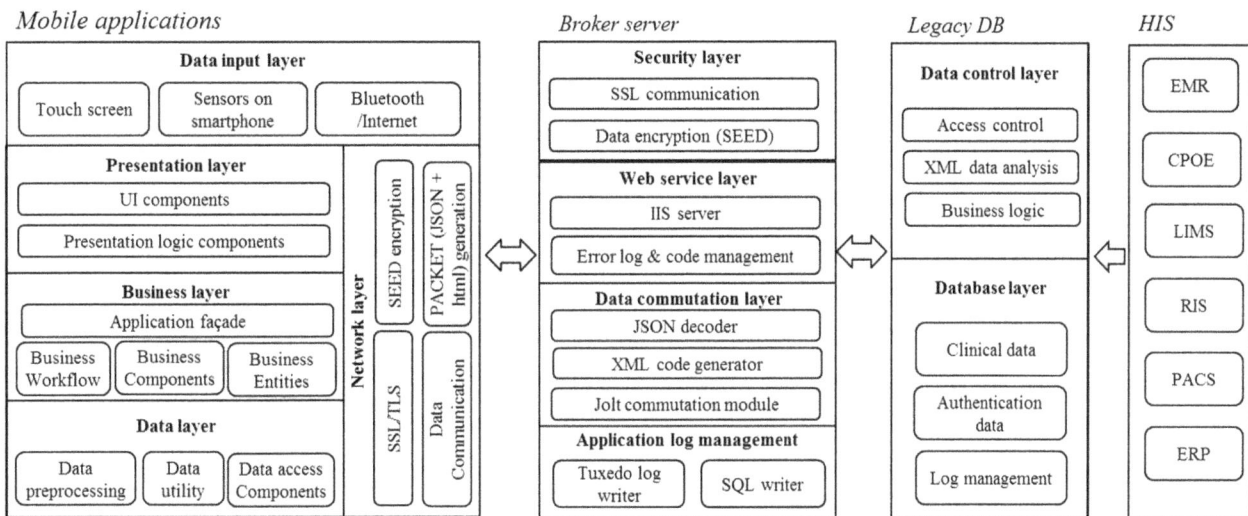

**Figure 5.** Technical architecture of mobile sensor applications. JSON: JavaScript Object Notation, SSL: Secure Socket Layer, TLS: Transport Layer Security, IIS: Internet Information Services, HIS: Hospital Information System, EMR: Electronic Medical Record, CPOE: Computerized Physician Order Entry, LIMS: Laboratory Information Management System, RIS: Radiology Information System, PACS: Picture Archiving Communication System, ERP: Enterprise Resource Planning.

## 3. Results and Discussion

### 3.1. My Cancer Diary

The "My Cancer Diary" app uses the smartphone camera for patients' log-in by recognizing the patient's barcode or QR code. This app was downloaded 7960 times via Google Play and installed in 2483 devices. The app was downloaded 2094 times via the iTunes App Store during this period (the iTunes App Store only provides the data since June 2013). Over 26 months, the sensor for "My Cancer Diary" was used 1194 times. The usual login by ID/password (PW) was used 56,151 times during the same period. Only 2% of logins used the barcode or QR code reader. Figure 6 shows the frequency of sensor usage by month with a trend line. The sensor usage by patients dramatically declined during the first three months. After six months, most patients shifted to use the conventional ID/PW method. We expected that an easy login using a camera scan would be beneficial for users; however, the results imply that the simple login capability with the camera scan only was not preferred by users.

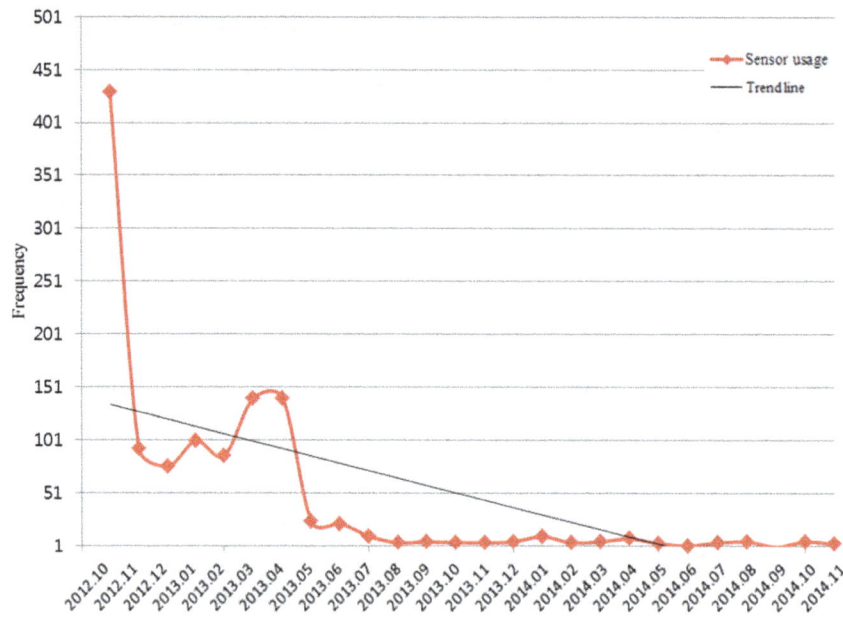

**Figure 6.** Frequency of sensor usage in "My Cancer Diary" by month.

*3.2. Point-of-Care HIV Check App*

The "Point-of-Care HIV Check" app identifies the patient's ID by a barcode and takes a photo of the test results using the smartphone camera. This app was released to 10 clinical nurse specialists in the emergency room [20]. This app was used 1565 times over 14 months. During this period, six HIV-infected patients were detected. Although this app demonstrated its benefits, the overall trend of usage declined slightly (refer Figure 7). The app was actively used during first six months; however, usage rapidly declined in April, 2014 when the charger of devices in the emergency room was out-of-order. This hardware problem was immediately resolved, but the app usage did not subsequently recover. Even though this app can provide a clear benefit to users, users regard the use of this app as an additional burden to their routine jobs.

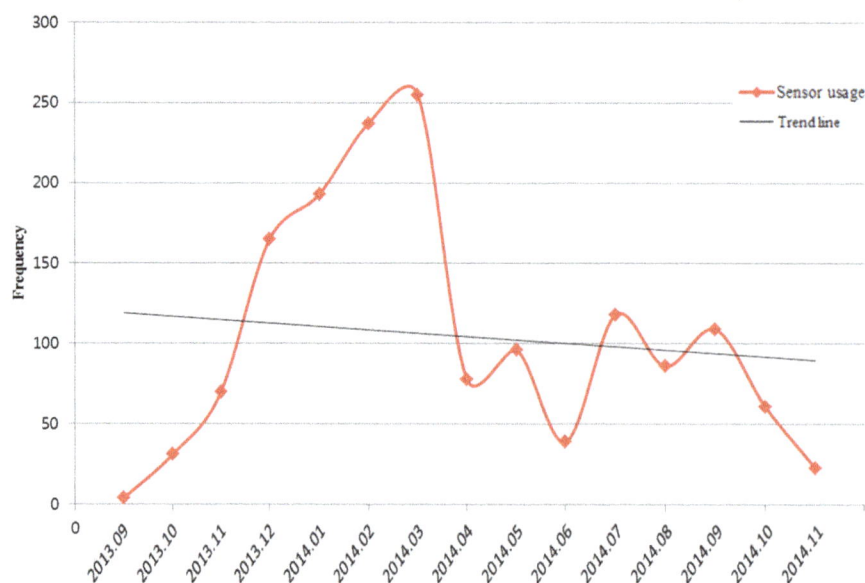

**Figure 7.** Usage pattern of Point-of-Care HIV check app by month.

## 3.3. Blood Culture App

The "Blood Culture" app identifies the patient's barcode using the iPhone camera and creates a timestamp for the blood culture. This app was released to four intern doctors at the pilot stages. And then it was released to all intern doctors in AMC (1568 persons). Up to now, 372 intern doctors have used this app. The app was used 3076 times over 29 months. When we looked at the log data of this app (Figure 8), there was one big peak and two minor peaks in the accumulated number of logins per month. One minor peak was in July 2012, around the pilot study of its practical usage. During the three-week pilot phase from 4 July 2012 to 26 July 2012 in the medical intensive care units, 356 sample data were collected by four intern doctors. Other peaks occurred from February 2013 to March 2013 and from February 2014 to March 2014. During February and March, most new trainees start their work rotation at the hospital. Considering the function of this application to guide the proper method for blood culture, this pattern of usage seems to be affected by the cycle of alterations of manpower and education for newcomers. However, the usage declined after one or two months.

When those interns evaluated the "Blood Culture" app, they evaluated this app as necessary for patient safety, patient and sample identification, and timeliness. Interestingly, one intern complained about the many data entry fields. However, this app was regarded as good for patient safety, identification, timeliness, and efficiency. This app could acquire important information about the sampler, sample time, sample site, and sample volume at the point of blood sampling without a delay from clinical processes.

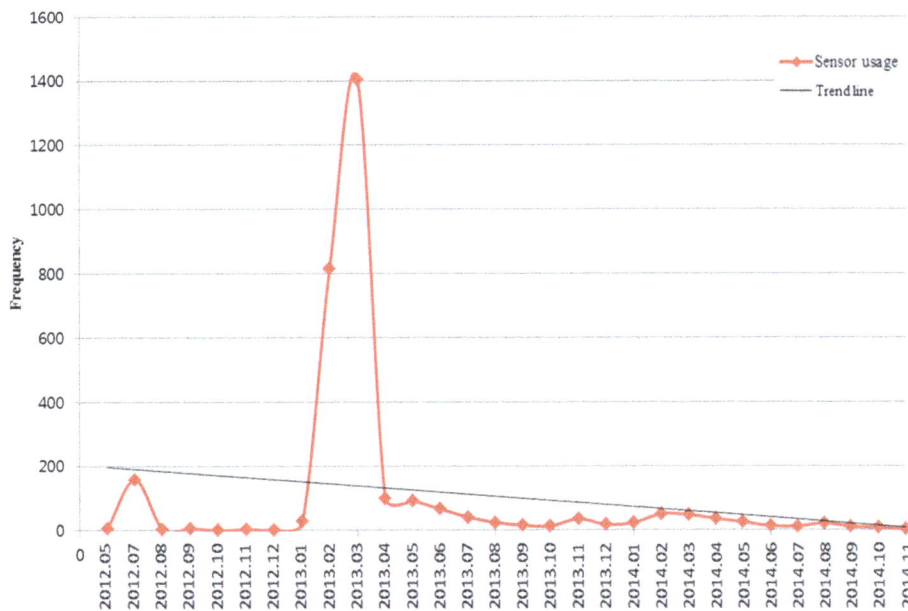

**Figure 8.** Frequency of usage of the "Blood Culture" app by month.

## 3.4. mAMIS

The mAMIS app was released to all physicians and nurses (5688 persons) in our hospital. Among them, 3004 persons used mAMIS. After the release of the upgraded mAMIS, the Touch ID login was used 3856 times during seven months. The usual ID/PW logins were used 250,129 times for the same period. Only 1.54% of logins on average used the Touch ID sensors. The rate looks still quite low compared to the ID/PW method. However, if we consider that only 38 users (1.3%) among total 3004 mAMIS users used Touch ID to login, the ratio seems to be reasonable. Interestingly, the usage pattern in Figure 9 steadily increased unlike those of the other apps analyzed. This implies that if mobile sensors can guarantee convenience or a workload reduction, users will voluntarily use mobile sensors.

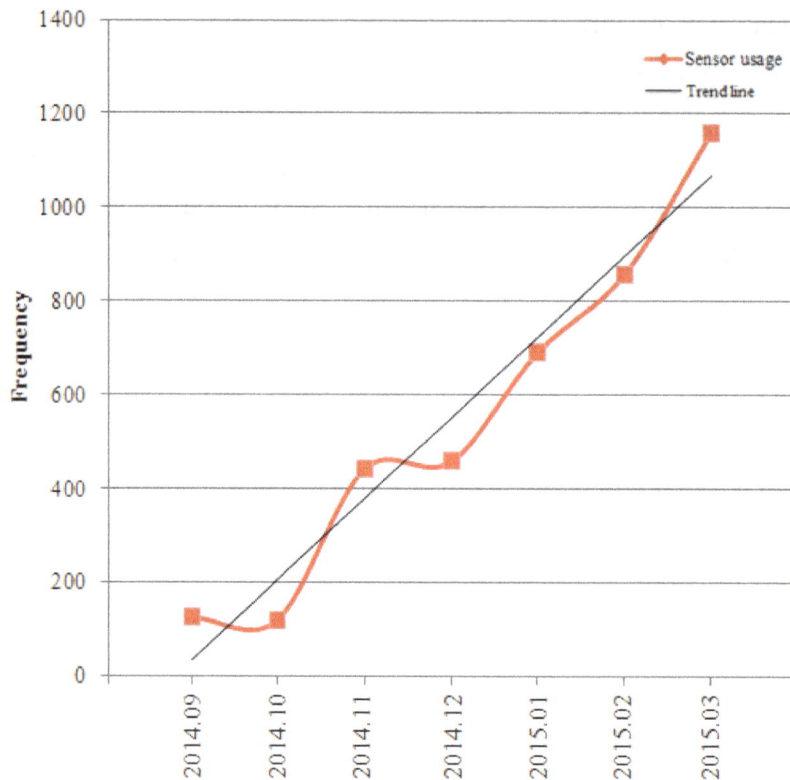

**Figure 9.** Frequency of usage in mAMIS app by month.

*3.5. Analysis of Usage Patterns of Healthcare Apps with Smartphone-Based Sensors*

Most of the four apps analyzed here were heavily used at the early stages of implementation or for some specific periods, but subsequently were not actively used. Though there could be many reasons for these declines in usages, and we suspect the following reasons: first, when launching new apps, the hospital actively promoted their usage by introducing them to users or by starting a pilot study. However, these promotion activities were suspended after a period of a few months. If the clinical activities or processes of user were not aligned with the usage of the developed apps, users started to reduce usage due to the additional steps required or the lack of a significant benefit to users as in the case of "Point-of-Care HIV Check" and "Blood Culture". This pattern implies that the usage of smartphone apps in a tertiary hospital is affected by the system or platform of the hospital, while other healthcare apps are chosen by personal demand. Hence, the apps with mobile sensors should be communicated or integrated into the hospital information system. Second, despite all the merits of smartphone apps with sensors, physicians may consider using them to be an additional skill or task that they have to practice in addition to an already demanding workload, not reducing the existing workload. Therefore, if this becomes inconvenient for some reason before users reach the plateau of the learning curve, the application might be abandoned [24]. The interaction with the user and updates of app functions at the AMC are not sufficient to address this inconvenience. As we can see in the case of "Point-of-Care HIV Check" app (Figure 7), a drop-off in usage was caused by a minor dysfunction of the devices or apps, and this could not be recovered. Third, users request conflicting features for healthcare apps as shown in Figure 10. Users want a device that is light-weight with a wider screen and operates quickly, but also supports sophisticated functions that usually run on a PC. Users also want to use the apps anywhere, anytime, and with a high security level. All these features exist in a trade-off relationship. That's why the usage rate of mobile apps for data entry drops quickly. At the early stage, users tried to use the data entry feature due to curiosity or the policy of the hospital. However, they reduced to use the feature because of those trade-offs.

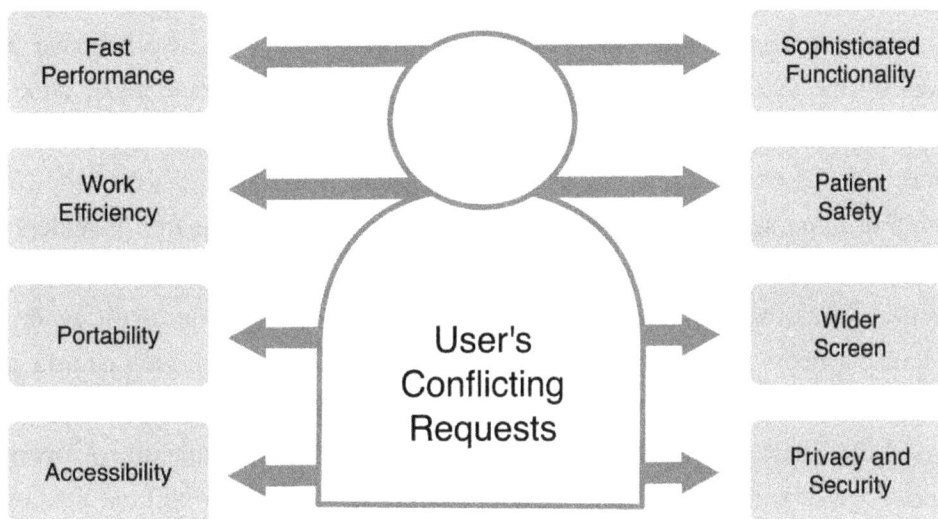

**Figure 10.** The conflicting requests of healthcare app users. Arrows indicate the trade-off relationships.

## 4. Conclusions and Outlook

Currently, the technology for mobile sensors in our hospital is far from mature and there are still technological issues to be addressed. These include: the inconvenience of mobile devices (small size of the window and difficulties entering characters), instability of the application, and low network speed and performance. There will also be challenges for building a data processing strategy in the hospital, because such sensor data represent a new type of information that has not been used before. The lack of a proper clinical procedure also needs to be resolved, such as requiring nurses to follow the laborious steps to use barcode system. Another major problem is the lack of a standardized format for sensor data. For instance, diverse barcode formats for patient identification are used in various departments. There also are potential security problems in the wireless network and applications that need to be tackled in relation to the deployment of mobile sensors in health.

However, despite any potential hurdles, it is likely only a matter of time before smartphone sensors are actively used in the hospital setting because healthcare apps with sensors can be used to monitor, warn, and provide health information. For example, the current mobile sensors, such as a portable pulse oximeter, have been used during the waiting time for examinations or at the time of transfer. However, the current portable pulse oximeters cannot record or send the data from patients to the EMR. Therefore, mobile devices connected to smartphones for monitoring vital signs that could be used universally by both expert and less experienced users, may reduce the blind spots in monitoring and recording [25]. In critically ill patient, those devices might provide a prompt alarm to the physician via the EMR system. Besides, if wireless devices for vital sign monitoring can be universalized, they may reduce the workload of health care providers to check and record vital signs manually. A smartphone with a camera can be a tool to record, monitor and communicate [26]. As we mentioned above, the smartphone camera can automatically send the images to the EMR system. In addition to the case in this paper, the photos of a skin lesion in the department of dermatology or the videos of neurologic examinations in the department of neurology may provide much more objective information than statements from physicians. However, when we use ordinary cameras for medical recording, it is troublesome not only because of its inconvenience, but also because of the security problems, caused by the use of memory cards to store the images. Because a QR code can be easily printed and quickly decoded, the QR code can be used for decision support system by containing the pharmacogenomic data of patients [27]. In a tertiary hospital, if we put the alert data of patients (e.g., bleeding tendency or hypersensitivity to specific drugs) into QR codes, health providers could receive critical information without needing to use a desk computer. A QR code alert system may serve to reduce the adverse events that happen during administration of drugs, examinations, or interventions, particularly, because sensors such as a barcode or QR code reader can provide fast and correct links to the EMR system. For this convenience, the main role of smartphone sensors may be as a bridge between users and the EMR system, especially during the periods of adaptation. However, QR codes have not been used actively in other industries [28]. Therefore, we need to find out the correct use cases of mobile sensors and educated the users to guarantee correct usage of sensors.

In addition, attachable sensors for smartphone can greatly expand the range of application [29,30]. As the size of medical devices has been getting smaller, the potential usage of smartphones will expand, not only as transmitters connecting to the EMR system, but also for a portable data-processing system. Traditionally, portable devices for diagnosis were regarded as preliminary methods before confirmative examination. However, palm-sized diagnostic devices are available now which have similar power as the enzyme-linked immunosorbent assay (ELISA), and can respond within fifteen minutes [29]. This allows smartphones with add-on sensors to provide medical care of higher quality to patients who cannot easily access a hospital, and even to the patients in unpredictable disasters [30]. In a tertiary hospital with more complex patient situations, these add-on sensors could increase the efficiency by increasing the discontinuation of the flow of medical care. Eventually, the cost of laboratory tests could be saved and the workload of healthcare providers might be reduced.

The potential of smartphones, amplified by various built-in and add-on sensors is limitless. Though there could be many trial-and-errors similar to our hospital' cases, smartphones with accompanying sensors may play their own distinct role in the medical field with their unique function and flexible usage. In this study, the graphs of frequency of usage against time were given to demonstrate the usage patterns. The development of the relevant metrics should be investigated to provide more insights into why the usage patterns decreased with time.

## Author Contributions

Yu Rang Park and Yura Lee analyzed the data and wrote the manuscript. Jae Ho Lee and Guna Lee wrote the manuscript. Soo-Yong Shin collected the data, wrote the manuscript and supervised the research.

## Conflicts of Interest

The authors declare no conflict of interest.

## References

1. Forrester Research. World Mobile and Smartphone Adoption Forecast, 2014 To 2015 (Global). Available online: https://www.forrester.com/Forrester+Research+World+Mobile+And+Smartphone+Adoption+Forecast+2014+To+2019+Global/fulltext/-/E-RES118252 (accessed on 15 December 2014).
2. Topol, E.J.; Steinhubl, S.R.; Torkamani, A. Digital medical tools and sensors. *JAMA* **2015**, *313*, 353–354.
3. Ozdalga, E.; Ozdalga, A.; Ahuja, N. The smartphone in medicine: A review of current and potential use among physicians and students. *J. Med. Internet Res.* **2012**, *14*, doi:10.2196/jmir.1994.
4. How are Physicians Using Mobile Apps for Professional Purposes? Available online: http://www.kantarmedia-healthcare.com/more-physicians-value-and-use-mobile-apps-for-professional-purposes (accessed on 12 December 2014).
5. Ventola, C.L. Mobile devices and apps for health care professionals: Uses and benefits. *P&T* **2014**, *39*, 356–364.
6. Global Mobile Health Market to Grow to $49B by 2020. Available online: http://mobihealthnews.com/30616/global-mobile-health-market-to-grow-to-49b-by-2020/ (accessed on 12 December 2014).
7. Emerging mHealth: Paths for Growth. Available online: http://www.pwc.com/gx/en/healthcare/mhealth/assets/pwc-emerging-mhealth-full.pdf (accessed on 12 December 2014).
8. Chung, E.H.; Guise, K.D. QTC intervals can be assessed with the AliveCor heart monitor in patients on dofetilide for atrial fibrillation. *J. Electrocardiol.* **2015**, *48*, 8–9.
9. Implanted Sensor Could Monitor Glaucoma and Prevent Blindness. Available online: http://mashable.com/2014/09/01/glaucoma-chip/ (accessed on 9 January 2015).
10. Examples of MMAs the FDA HAS Cleared or Approved. Available online: http://www.fda.gov/MedicalDevices/ProductsandMedicalProcedures/ConnectedHealth/MobileMedicalApplications/ucm368784.htm (accessed on 15 December 2014).
11. Bennet, Hatch Introduce Bill to Cut Red Tape, Boost Innovation in Health IT. Available online: http://www.hatch.senate.gov/public/index.cfm/releases?ID=750a5957-08e0-4a77-95ba-5806a7f3ffa7 (accessed on 15 December 2014).
12. Did You Know How Many Different Kinds of Sensors Go Inside a Smartphone? Available online: http://www.phonearena.com/news/Did-you-know-how-many-different-kinds-of-sensors-go-inside-a-smartphone_id57885 (accessed on 20 January 2015).

13. Sebrook, H.J.; Stromer, J.N.; Shevkenek, C.; Bharwani, A.; de Grood, J.; Ghali, W.A. Medical applications: A database and characterization of apps in Apple iOS and Android platforms. *BMC Res. Notes* **2014**, *7*, doi:10.1186/1756-0500-7-573.

14. Ryu, H.J.; Kim, W.S.; Lee, J.H.; Min, S.W.; Kim, S.J.; Lee, Y.H.; Nam, S.W.; Eo, G.S.; Seo S.G.; Nam, M.H. Asan medical information system for healthcare quality improvement. *Healthc. Inf. Res.* **2010**, *16*, 191–197.

15. Park, J.-Y.; Lee, G.; Shin, S.-Y.; Kim, J.H.; Han, H.-W.; Kwon, T.-W.; Kim W.S.; Lee, J.H. Lessons learned from the development of health applications in a tertiary hospital. *Telemed. J. E Health* **2014**, *20*, 215–222.

16. Pandey, A.; Hasan, S.; Dubey, D.; Sarangi, S. Smartphone apps as a source of cancer information: Changing trends in health information-seeking behavior. *J. Cancer Educ.* **2013**, *28*, 138–142.

17. Porta, C.; Handelman, E.; McGovern, P. Needlestick injuries among health care workers: A literature review. *AAOHN J.* **1999**, *47*, 237–244.

18. Trim, J.C.; Elliott, T.S. A review of sharps injuries and preventative strategies. *J. Hosp. Infect.* **2003**, *53*, 237–242.

19. Do, A.N.; Ciesielski, C.A.; Metler, R.P.; Hammett, T.A.; Li, J.; Fleming, P.L. Occupationally acquired human immunodeficiency virus (HIV) infection: national case surveillance data during 20 years of the HIV epidemic in the United States. *Infect. Control Hosp. Epidemiol.* **2003**, *24*, 86–96.

20. Jang, D.; Shin, S.-Y.; Seo, D.-W.; Joo, S.; Huh, S.-J. A Smartphone-based system for the automated management of point of care test results in hospitals. *Telemed. J. E Health* **2015**, *21*, 301–305.

21. Karchmer, A.W. Infective Endocarditis. In *Harrison's Principles of Internal Medicine*, 18th ed.; Longo, D.L., Fauci, A.S., Kasper, D.L., Hauser, S.L., Jameson, J., Loscalzo, J., Eds.; McGraw-Hill: New York, NY, USA, 2012; pp. 1056–1057.

22. Lee, J.; Chong, Y.; Jang, S.; Kim, M.; Lee, G.; Kim, J.; Kwon, T.; Kim, W. Development of Smartphone Blood Culture Application Using Barcode and Hospital Information System: A University Hospital Experience. In Proceedings of the American Medical Informatics Association Annual Symposium, Washington, DC, USA, 16–20 November 2013; p. 851.

23. Introducing Oracle Jolt. Available online: http://docs.oracle.com/cd/E13161_01/tuxedo/docs10gr3/jdg/dvintro.html (accessed on 10 April 2015).

24. Ritter, F.E.; Schooler, L.J. The learning curve. In *International Encyclopedia of the Social and Behavioral Sciences*; Pergamon: Amsterdam, The Netherlands, 2002; pp. 8602–8605.

25. Hudson, J.; Nguku, S.M.; Sleiman, J.; Karlen, W.; Dumont, G.A.; Petersen, C.L.; Warriner, C.B.; Ansermino, J.M. Usability testing of a prototype Phone Oximeter with healthcare providers in high- and low-medical resource environments. *Anaesthesia* **2012**, *67*, 957–967.

26. Newmark, J.L.; Ahn, Y.K.; Adams, M.C.; Bittner, E.A.; Wilcox, S.R. Use of Video Laryngoscopy and Camera Phones to Communicate Progression of Laryngeal Edema in Assessing for Extubation: A Case Series. *J. Intensive Care Med.* **2013**, *28*, 67–71.

27. Minarro-Gimenez, J.A.; Blagec, K.; Boyce, R.D.; Adlassnig, K.P.; Samwald, M. An ontology-based, mobile-optimized system for pharmacogenomic decision support at the point-of-care. *PLoS ONE* **2014**, *9*, doi:10.1371/journal.pone.0093769.

28. Are QR Codes Dead? Available online: http://www.forbes.com/sites/ilyapozin/2012/03/08/are-qr-codes-dead/ (accessed on 10 April 2015).

29. Laksanasopin, T.; Guo, T.W.; Nayak, S.; Sridhara, A.A.; Xie, S.; Olowookere, O.O.; Cadinu, P.; Meng, F.; Chee, N.H.; Kim, J.; *et al.* A smartphone dongle for diagnosis of infectious diseases at the point of care. *Sci. Transl. Med.* **2015**, *7*, doi:10.1126/scitranslmed.aaa0056.

30. Adhikari, S.; Blaivas, M.; Lyon, M.; Shiver, S. Transfer of real-time ultrasound video of FAST examinations from a simulated disaster scene via a mobile phone. *Prehosp. Disaster Med.* **2014**, *29*, 290–293.

# 5

# Impedance of the Grape Berry Cuticle as a Novel Phenotypic Trait to Estimate Resistance to *Botrytis Cinerea*

**Katja Herzog \*, Rolf Wind and Reinhard Töpfer**

Julius Kühn-Institut-Federal Research Centre of Cultivated Plants, Institute for Grapevine Breeding Geilweilerhof, Siebeldingen 76833, Germany; E-Mails: rolf.wind@kdwelt.de (R.W.); reinhard.toepfer@jki.bund.de (R.T.)

\*  Author to whom correspondence should be addressed; E-Mail: Katja.herzog@jki.bund.de

Academic Editor: Gonzalo Pajares Martinsanz

**Abstract:** Warm and moist weather conditions during berry ripening provoke *Botrytis cinerea* (*B. cinerea*) causing notable bunch rot on susceptible grapevines with the effect of reduced yield and wine quality. Resistance donors of genetic loci to increase *B. cinerea* resistance are widely unknown. Promising traits of resistance are represented by physical features like the thickness and permeability of the grape berry cuticle. Sensor-based phenotyping methods or genetic markers are rare for such traits. In the present study, the simple-to-handle I-sensor was developed. The sensor enables the fast and reliable measurement of electrical impedance of the grape berry cuticles and its epicuticular waxes (CW). Statistical experiments revealed highly significant correlations between relative impedance of CW and the resistance of grapevines to *B. cinerea*. Thus, the relative impedance $Z_{rel}$ of CW was identified as the most important phenotypic factor with regard to the prediction of grapevine resistance to *B. cinerea*. An ordinal logistic regression analysis revealed a $R^2_{McFadden}$ of 0.37 and confirmed the application of $Z_{rel}$ of CW for the prediction of bunch infection and in this way as novel phenotyping trait. Applying the I-sensor, a preliminary QTL region was identified indicating that the novel phenotypic trait is as well a valuable tool for genetic analyses.

**Keywords:** sensor development; phenotyping; grapevine breeding; berry skin; objective data; bunch compactness; *Vitis vinifera*

## 1. Introduction

Grey mold is a plant disease caused by the ubiquitous fungus widely known as Botrytis that affects more than 200 plant species [1]. The necrotrophic pathogen and filamentous fungus *Botrytis cinerea* PERS., abbreviation *B. cinerea*, is the anamorph of the ascomycete *Botryotinia fuckeliana* WHETZEL. On grapevine (*Vitis vinifera* L.) it causes one of the most serious diseases, the bunch rot. This disease can drastically reduce both the yield at harvest time and wine quality, and can be controlled by specific canopy management, *i.e.*, the reduction of foliage around grape bunches (literature overview is given by Molitor, *et al.* [2], Broome, *et al.* [3]) permitting a faster drying of grape bunches. In years with persistent rain during the ripening period, the effectiveness of canopy management is limited and expensive fungicide applications [4] or, at the expenses of quality, a premature harvest is necessary in order to keep yield losses at a minimum. Furthermore, bunch rot can be observed especially on grapevines with compact bunch architecture [2,5–8], as illustrated in Figure 1, and occurrence of the disease is most notable in years with moist and warm weather conditions during ripening of the grape berry [3,9,10]. With regard to that, in grapevine breeding programs, seedlings will be selected with convenient physical properties, e.g., loose bunch architecture and small berries.

**Figure 1.** The compactness of grapevine bunches as one major reason for the susceptibility of grapevines to *B. cinerea*. Compact bunches of the susceptible grapevine cultivar 'Riesling' (**a**); showing regions where berries are very dense (**b**); and often the growth of *B. cinerea* begins in these regions as result of damaged berries (**c**).

Besides compactness, different berry skin features seem to influence the susceptibility of grapevines towards *B. cinerea* infection, *i.e.*, the biochemical composition [10–13], the ripening stage [10,12] and the morphology of the berry skin [11,14]. Especially, the cuticle and its epicuticular waxes are described as important berry skin features regarding the susceptibility of berries toward bunch rot [11,15,16]. In this context, warm temperatures, high air humidity and water on the berry surface are known as major reasons for the incidence of microscopic cracks in the cuticle membrane of berries [15,16]. These, in turn, play a critical role in the susceptibility of grape berry against *B. cinerea*, since they impair the function of the cuticle as a barrier for pathogen defense and permit an increased water uptake into the berries [16]. The cuticle forms the outer surface of leaf, fruit and primary-shoot epidermal cell walls [17,18] and additionally serves as a regulator of molecular diffusion [19]. It consists of intra- and epicuticular waxes, which build up a hydrophobic berry surface. That, in turn, achieves a faster drying of berries/bunches, which is described as an important factor in reducing the susceptibility of grapevines to *B. cinerea*. It is also known that polar pores appear through high air humidity and warm conditions resulting in an increased water permeability of this pores [20,21]. Polar pores in return facilitate the diffusion of organic substrates (*i.e.*, sugar, nutrients) to the berry surface and promoting growth of *B. cinerea*. Figure 2 illustrates the impact of the thickness of cuticle and epicuticular waxes on: (1) transport of nutrients to the berry surface; (2) hydrophobic property on the berry surface; and (3) accumulation of water between berries of compact bunches.

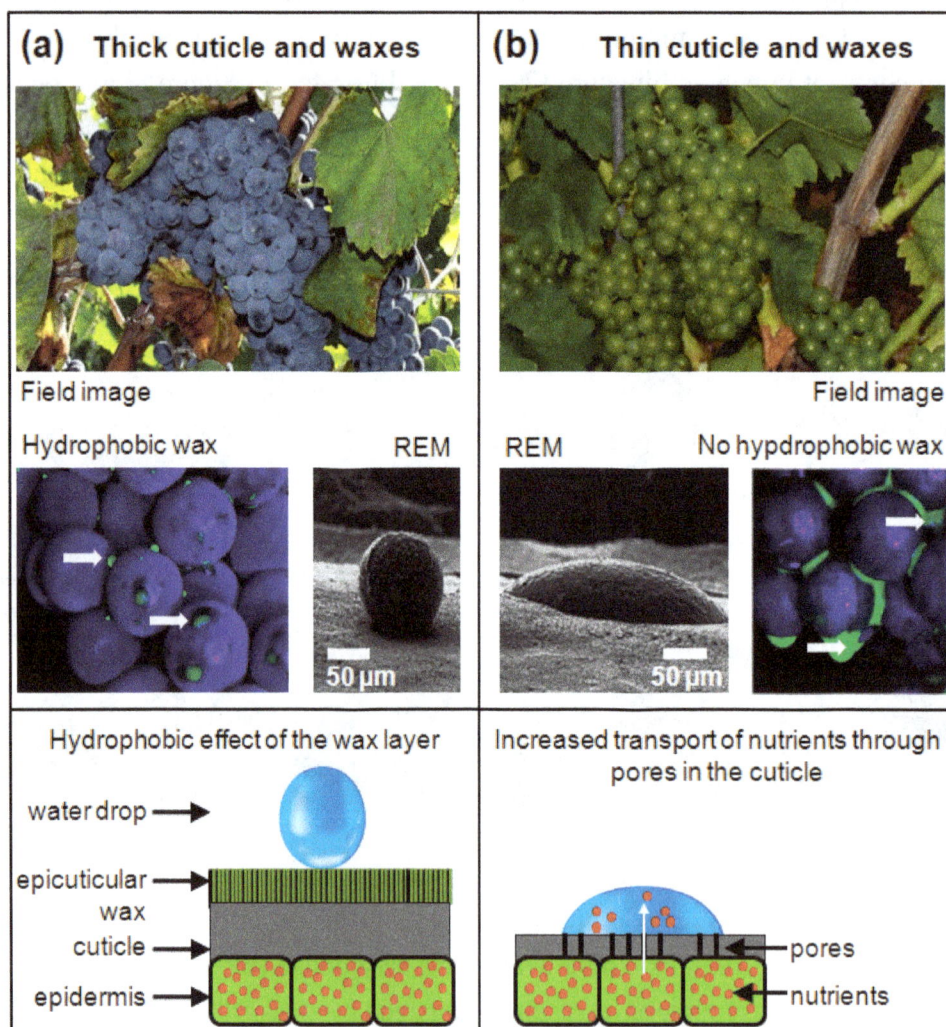

**Figure 2.** Function of the cuticle and epicuticular waxes as physical barriers. (**a**) As an example of grape berries with a thick cuticle and waxes, the grapevine accessions Seibel 182 from the genetic repository in Siebeldingen was used. The hydrophobic characteristic of epicuticular waxes (wax layer) permits fast drying of berry surfaces; (**b**) As an example of grape berries with thin cuticle and waxes the grapevine cultivar 'Morio Muskat' is shown. The absence of epicuticular waxes results in an accumulation of water between berries. Fluorescein (yellow-green) stained water (arrows) as well as Raster Electron Microscope (REM) were used to illustrating the effect.

The current bottleneck in phenotyping physical characteristics of berries is the lack of an easy and reliable method. The phenotyping methods described in previous studies (e.g., phenotyping the thickness of the epidermal layer or number of pores or existence of microscopic cracks) are laborious and very time consuming [11,15,16]. Hence, the acquisition of these traits is not feasible for common grapevine breeding programs where hundreds of different genotypes need to be evaluated. Therefore, the development of sensor-based methods is required in order to phenotype the grape berry skin and the influence to the bunch rot susceptibility of grapevines. The method should be rapid, permitting the generation of objective phenotypic data from a large number of samples. With regard to this, the measurement of electrical impedance $Z$ was chosen to characterize electrical behavior of the berry cuticle (C) and epicuticular waxes (W) as a novel phenotypic trait as well as an indicator for thickness and permeability of C and W (Figure 3).

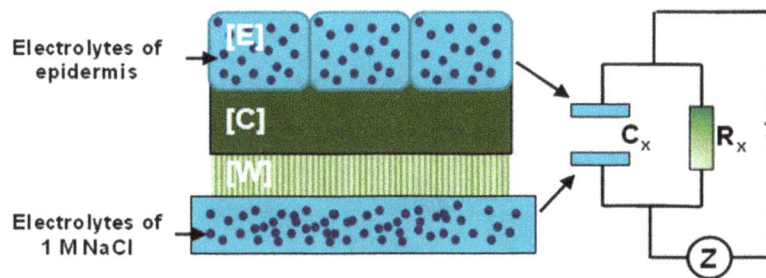

**Figure 3.** Schematic distribution of electrolytes and the physical principle of impedance measurements. The cuticle [C] and the epicuticular wax layer [W] located between two electrically conducting surfaces, the epidermis [E] and NaCl solution. The impedance $Z$ is the sum of the imaginary resistor ($C_x$ (thickness of cuticle, wax layer and air)) and the real resistor ($R_x$ (permeability of [C] and [W]).

The first aim of the present study was the development and validation of a sensor to determine impedance of the cuticle of grape berries with epicuticular waxes ($Z_{rel}$ of CW) and the cuticle without epicuticular waxes ($Z_{rel}$ of C). Secondly, both physical parameters were measured from different grapevine cultivars to investigate if there is any relationship between impedance and the susceptibility of grape berry toward *B. cinerea* infection and bunch rot, respectively. Finally, as a test of the application, the impedance was measured in a F1 progeny (crossing of GF.GA.47-42 x 'Villard Blanc') in order to test the novel phenotypic data for its utilization in QTL (Quantitative-Trait-Loci) analysis.

## 2. Experimental Section

### 2.1. Plant Material and Sampling

As plant material, 41 different genotypes (including traditional cultivars, breeding material and cultivars from the genetic repository planted at the experimental vineyard of Geilweilerhof located in Siebeldingen, Germany ((N 49°21.747, E 8°04.678), overview in Table S1) were used for method validation. In addition, 144 genotypes of a F1 progeny of the crossing GF.GA-47-42 (crossing of 'Bacchus Weiss' x 'Seyval') x 'Villard Blanc' (crossing of 'Seibel 6468' x 'Subereux') were used for phenotyping and QTL analysis. All investigated grapevines are planted in North-South orientation in the vineyards at Geilweilerhof.

Berry ripening for each genotype was monitored weekly by the measurement of sugar content applying a handheld refractometer (VWR® International GmbH, Darmstadt, Germany). Once the sugar level of berries reached approximately 70° Oechsle (*i.e.*, 17.1% Brix), two bunches were sampled (one from the east and the other one from the west side of the plant). Fifteen visually intact berries were randomly sampled per bunch (30 berries per cultivar) by cutting them off carefully at the berry pedicel.

## 2.2. Construction of the I-Sensor

Prototype I-sensor was developed to acquire the electrical impedance of the grape berry cuticle and epicuticular waxes (Figure 4).

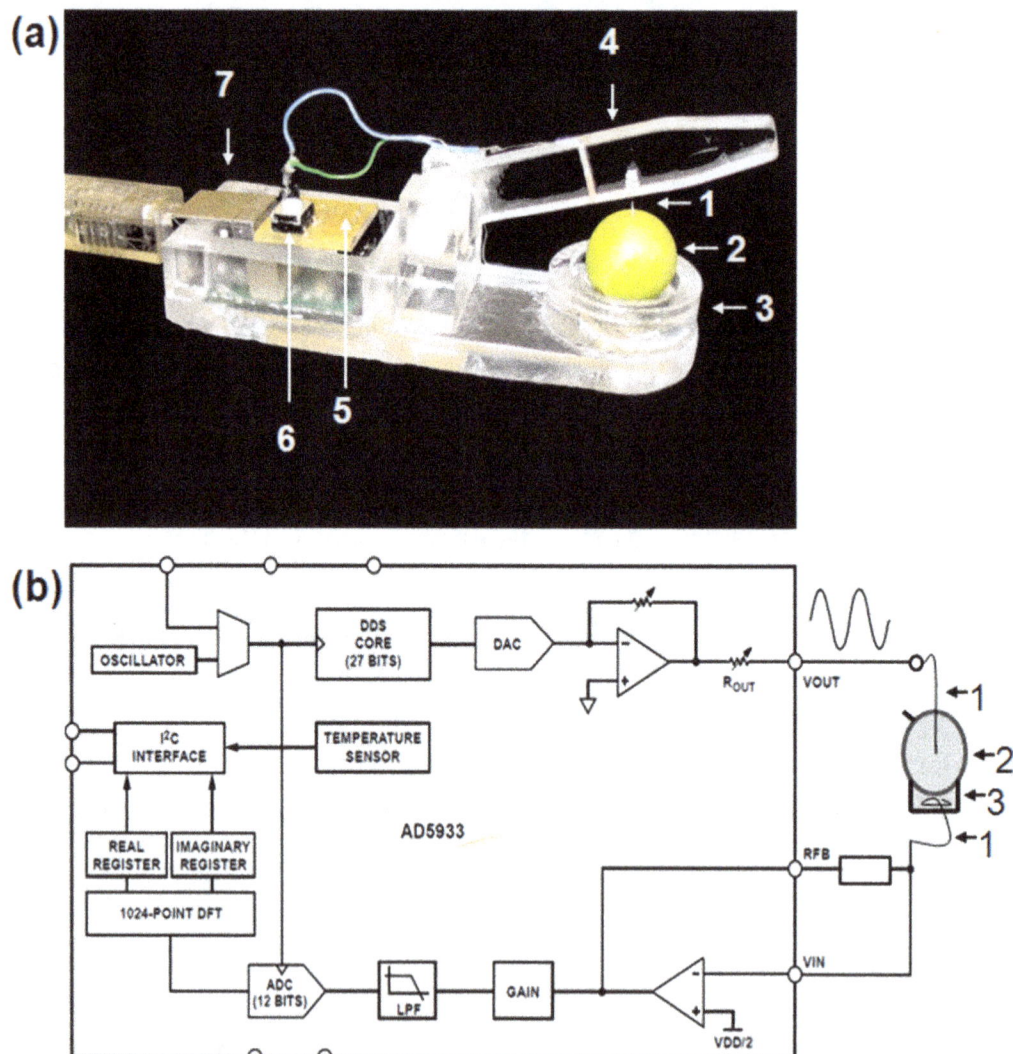

**Figure 4.** Setup (**a**) and functional block diagram (**b**) of the I-sensor. 1 = sensing electrodes; 2 = grape berry; 3 = skip for NaCl solution; 4 = mobile sensor head; 5 = AD5933 impedance converter; 6 = Button for sensor calibration; and 7 = USB-I²C module.

The AD5933 high precision impedance converter system (Analog Diveces GmbH, Munich, Germany) fused with a USB-I$^2$C module (Devantech Ltd (Robot Electronics, Norfolk, United Kingdom), which provides a complete interface between PC and the I$^2$C (*Inter Integrated Circuit*) bus. The USB-I$^2$C module uses the FTDI FT232R USB chip. Platinum-Iridium wires with a diameter of 0.3 mm were used as sensing electrodes. For sensor operating, *i.e.*, configuration, calibration, measurement, and result export, a software tool was developed using Embarcadero Delphi Version 3 (Borland$^®$, Austin, TX, USA).

## 2.3. Impedance Measurements

For the measurement of the impedance $Z$, single grape berries were placed in the I-sensor (component 3 in Figure 4a) containing a 1 M sodium chloride (NaCl) solution. The mobile sensor head was used to prick the Platinum-Iridium wire (sensing electrode) into the berry. Measurements were conducted at room temperature using an electrical frequency of 2 KHz and 30 KHz. The lower frequency of 2 KHz represents the permeability of the cuticle (C) and cuticle with epicuticular waxes (CW) because it is closely related to the direct current (DC), which is used to determine the threshold voltage. In addition, the higher frequency of 30 KHz represents the thickness of C and CW. The acquisition of $Z$ of CW was carried out by measuring visual intact berries twice. Afterwards, the epicuticular waxes were mechanically removed by carefully rubbing using Kimtech-Science$^®$ Precision wipes (Kimberly-Clark$^®$ Professional, Kimberly-Clark GmbH, Koblenz-Rheinhafen, Germany). The mechanical removal of the wax layer was the most practical, cheapest and fastest way and ensured that the cuticle or other berry skin components were not changed in its chemical or physical properties as it would be the case by using chemicals for eliminating the wax layer. The measurements were repeated to acquire $Z$ of C. For both, two different berry positions (lateral and bottom) from the 30 sampled berries were determined, *i.e.*, 60 impedance measurements per genotype and treatment. Subsequently, the berries were bisected and the basis impedance $Z_B$ (impedance of the berry flesh) was determined.

The impedance $Z$ at 2 KHz and 30 KHz were used to calculate the relative impedance $Z_{rel}$, whereby $d$ is the difference between the used electrical frequency ($d = 28$)

$$Zrel = 0.5 \times (Z_{2KHz} + Z_{30KHz}) \times d \times 0.001 \tag{1}$$

For further investigations, the median of the 60 $Z_{rel}$ values (minus $Z_B$) was calculated for each genotype.

## 2.4. QTL Analysis

The median of the relative impedance was determined from berries of 144 plants of the F1 progeny (crossing of GF.GA-47-42 x 'Villard Blanc'). The population and a first map is described by Zyprian, *et al.* [22] and was extended with additional markers [23,24]. In the present study this extended map was used for QTL analysis. QTL analysis was carried out applying MapQTL$^®$ 6.0 (Kyazma$^®$, Wageningen, The Netherlands) as described by Fechter, *et al.* [23]. Interval mapping (IM) and permutation test were used to identify preliminary QTLs whose flanking markers were used as co-factors for multiple QTL mapping (MQM).

## 2.5. Reference Evaluations

The grapevine phenology was evaluated using the BBCH scale (Biologische Bundesanstalt, Bundessortenamt und Chemische Industrie) [25]. It is a commonly used evaluation system to describe predefined stages of plant development. The compactness of grape bunches was classified in parallel to berry sampling by using the OIV (International Organization of Vine and Wine) descriptor 204 [26]. Therefore, the bunch compactness was estimated for all bunches of numerous plants when bunches were sampled for impedance measurements. The susceptibility of genotypes to *B. cinerea* was classified under natural field conditions three weeks after the measurement of impedance. The OIV descriptor 459 was used to create a modified five-class scale (Figure 5).

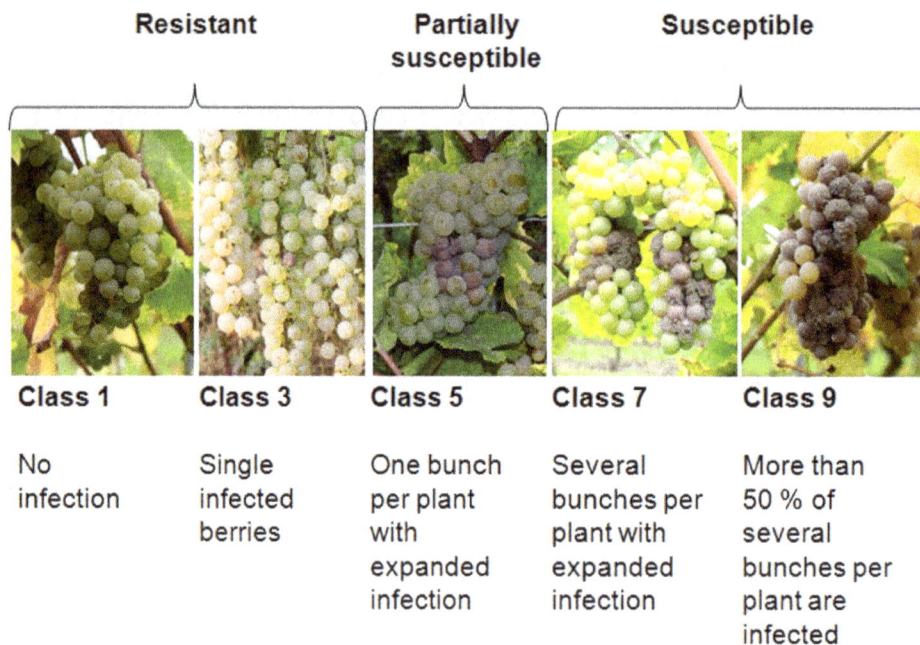

| Resistant | | Partially susceptible | Susceptible | |

| Class 1 | Class 3 | Class 5 | Class 7 | Class 9 |
|---|---|---|---|---|
| No infection | Single infected berries | One bunch per plant with expanded infection | Several bunches per plant with expanded infection | More than 50 % of several bunches per plant are infected |

**Figure 5.** Classification of grapevine susceptibility to *B. cinerea* infection.

The aim was the estimation of the risk for *B. cinerea* infection. Genotypes classified in class 3 show only single *B. cinerea* infected berries. These grapevines were denoted as resistant. When several bunches showed many *B. cinerea* infected berries, they were classified as class 3–5, and when the infection expands (example in Figure 1) they were classified as class 5 (*i.e.*, partially infected). This is the beginning of spread of the *B. cinerea* infection, which could be a problem during ripening especially in times of persistent rain. When the infection has expanded on several bunches the grapevines were classified as susceptible.

It was assumed that genotypes that were classified into: class 1 and 3 are resistant; class 5 are partially susceptible; and class 7 and 9 represent susceptible genotypes.

## 2.6. Statistical Analysis

For method validation, different statistical analyses were conducted using the software SAS® (Statistical Analysis System) Enterprise Guide 4.3 (SAS Institute Inc., Cary, NC, USA). The mean relative impedance $Z_{rel}$ of investigated berries was compared with *B. cinerea* resistance of the regarding genotype, which had been evaluated in the field. One-way ANOVA analysis, *Pearson* correlation coefficient and *Duncan* multiple range test with the level of significance $\alpha = 0.05$ were conducted. Logistic regression analysis (PROC LOGISTIC) was performed. For the prediction of the probability of *B. cinerea* infection (Five class classification) Maximum Likelihood estimation was used. McFadden's pseudo coefficient of determination $R^2$ ($R^2_{McFadden}$) was calculated utilizing $-2$ LOG L (*i.e.*, the log Likelihood of the fitted model) of L0 (constant model) and L1 (constant model and covariates).

$$R^2_{\text{McFadden}} = 1 - \left(\frac{L_1}{L_0}\right) \tag{2}$$

## 3. Results and Discussion

### 3.1. Validation the Functionality of the I-Sensor

For the validation of the I-sensor measurements, the impedance was determined 50-times from five visually intact berries of three grapevine cultivars. Table S1 shows the mean impedance for the three cultivars, indicating significant differences between genotypes. The data obtained by the I-sensor proved to be highly reproducible with low standard deviation (SD). Thus, the instrument was usable in a novel phenotyping approach to characterize grape berry cuticle and epicuticular waxes.

### 3.2. Novel Phenotypic Trait as Indicator for Resistance of Grapevines to B. cinerea

The *Pearson* correlation coefficient (Table 1) and significance of correlation were calculated to consider the relation of the mean $Z_{rel}$ values from 40 grapevine genotypes (Figure S1) with the corresponding class of susceptibility to *B. cinerea*.

**Table 1.** *Pearson* correlation coefficients and significance of correlation of relative impedance $Z_{rel}$ and bunch compactness determined in comparison to the evaluated *B. cinerea* susceptibility of grapevine cultivars in the field. CW: intact cuticle with epicuticular waxes; C: cuticle without epicuticular waxes; W: epicuticular wax calculated by the substraction of $Z_{rel}$ of CW and $Z_{rel}$ of C.

| $Z_{rel}$ | Architecture | Remark | Susceptibility to *B. cinerea* | Significance |
|---|---|---|---|---|
| CW | | $Z_{rel}$ of CW | **−0.67** | <0.0001 |
| C | | $Z_{rel}$ C | −0.60 | <0.0001 |
| W | | $Z_{rel}$ of CW − $Z_{rel}$ of C | −0.53 | 0.0004 |
| | Bunch compactness | | 0.40 | 0.0096 |

The highest negative correlation was detected between $Z_{rel}$ of CW (the cuticle of grape berry with epicuticular waxes) and *B. cinerea* susceptibility. This result indicates the importance of both berry skin features with regard to the mechanical protection towards *B. cinerea*. It was observed that genotypes revealing impedance values of CW of 600 or greater show high resistance to *B.cinerea*. In contrast to the literature [2,5–8], in the present study, the bunch compactness showed only a minor positive correlation to the infection of bunches with *B. cinerea*. This result revealed that the impedance of the berry cuticle and its waxes plays an important role with respect to the susceptibility of investigated cultivars to bunch rot.

To normalize the phenotypic data, the data set was grouped depending on the evaluated grape bunch compactness, *i.e.*, loose, medium or compact bunches. Again, the *Pearson* correlation coefficients and significance of correlations were calculated for each group (Table 2).

**Table 2.** *Pearson* correlation coefficient and significance of correlations for the susceptibility of grapevines to *B. cinerea* and the mean $Z_{rel}$ of CW, C and W. Grapevine genotypes were grouped according to their bunch compactness, which was evaluated using OIV descriptor 224. The correlation coefficient was rated according to Bühl [27]. N = number of samples; CW: intact cuticle with epicuticular waxes; C: cuticle without epicuticular waxes; W: epicuticular wax.

| Bunch Compactness | N | $Z_{rel}$ | Susceptibility to *B. cinerea* | Significance | Rating of Correlation |
|---|---|---|---|---|---|
| loose | 10 | CW | −0.43 | n.s. | Low |
| | | C | −0.22 | n.s. | Low |
| | | W | −0.50 | n.s. | Low |
| medium | 17 | CW | **−0.72** | 0.001 | **High** |
| | | C | −0.57 | 0.0173 | Medium |
| | | W | −0.61 | 0.01 | Medium |
| compact | 21 | CW | **−0.80** | <0.0001 | **High** |
| | | C | **−0.83** | <0.0001 | **High** |
| | | W | −0.62 | 0.0027 | Medium |

It was discovered that the susceptibility to *B. cinerea* is significantly correlated with the impedance $Z_{rel}$ when the bunch compactness was medium or high. Presumably, the physical property of loose bunches lead to a lower *B. cinerea* infection risk. In loose bunches, the berries do not touch each other. It is considered that the contacts between berries results in violation of the cuticle, *i.e.*, microscopic cracks emerge and *B. cinerea* can easily penetrate the berries [15]. Furthermore, the accumulation of water between berries (slow drying of the berry surface) is reduced in loose bunches, which also restricts the appearance of bunch rot. For further statistical analysis, it was thus assumed that for grapevines with loose bunches, resistance to *B. cinerea* was not mainly influenced by the properties of cuticle, but the bunch architecture itself. The data from these grapevine genotypes were not considered in the regression analysis. In contrast, the *Pearson* correlation coefficients of the group of medium bunch compactness (Table 2) indicate a high negative relation between the impedance $Z_{rel}$ of CW and *B. cinerea* susceptibility. This relation is even 10% higher within the group of compact bunches. Hereby, the architecture of grape bunches is supposed as one major explanation. The correlation data in Table 2 indicate that the impedance of the cuticle is more correlated with *B. cinerea* resistance within the compact group than in the medium group. For the investigated genotypes, it could be assumed that genotypes with thick berry cuticles (high impedance) are more resistant than genotypes with thin cuticles (low impedance). In compact bunches, the berries touch each other (Figure 1), which causes injury to berry skin and subsequently promotes *B. cinerea* infection. It seems that the physical properties of the berry skin (thickness and permeability) in the group of compact bunches is more related to *B. cinerea* infection than in the group of medium bunches. This consideration indicates an overlay of bunch architecture and impedance of the cuticle and epicuticular waxes in relation to bunch rot.

To sum up, the $Z_{rel}$ of CW ($Z_{rel}$ of the cuticle with epicuticular waxes) showed the highest significant correlations. Thus, it was selected as the most promising indicator for the prediction of resistance of grape berries to *B. cinerea* using an ordinal regression analysis. However, the investigated genotypes were grouped on the basis of their corresponding *B. cinerea* susceptibility, *i.e.*, a resistant, partially susceptible, and susceptible group, and with regard to the level of bunch compactness (medium and compact). The mean impedance $Z_{rel}$ of CW with standard deviation was calculated for each group (Figure 6).

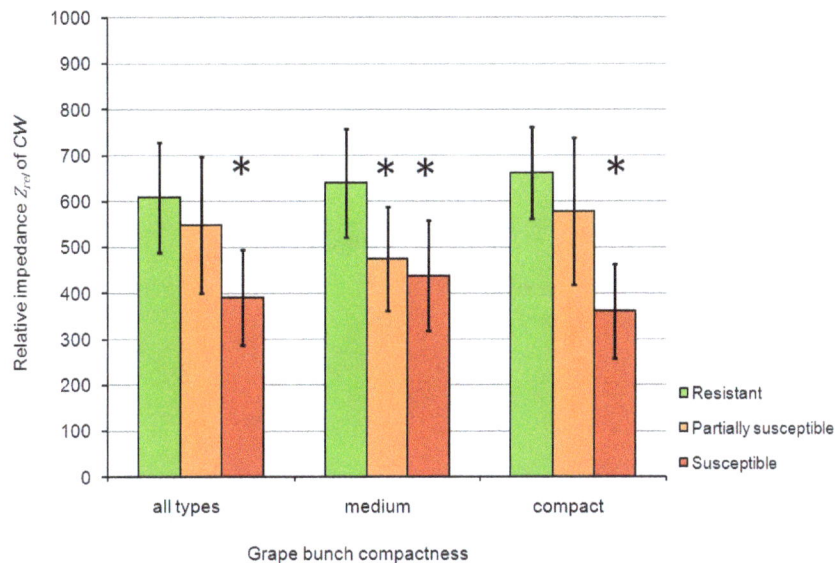

**Figure 6.** Mean relative impedance $Z_{rel}$ of CW of three different groups: resistant, partially susceptible and susceptible grapevines. The Duncan multiple range test (level of significance $\alpha = 0.05$) was carried out to distinguish the groups statistically based on the mean impedance. Standard deviation is shown in the graphs. Stars indicate significant differences in comparison to the resistant group. CW: intact cuticle with epicuticular waxes.

Significant differences of $Z_{rel}$ of CW were identified between the resistant and susceptible groups (Figure 6). No significant differences were observed between the average impedance of the resistant and partially susceptible group (all types of grape bunch compactness and compact bunches). Therefore, it could be helpful to investigate an enlarged set of cultivars. As a consequence, the distinction between susceptible and resistant grapevines is principally possible using the impedance value as a novel type of phenotypic data.

Logistic regression analysis was carried out in order to predict the probability of *B. cinerea* infection by using the relative impedance $Z_{rel}$ of CW (Figure 7).

In the regression analysis, an $R^2_{McFadden}$ of 0.37 was calculated. As visible in Figure 7, the classes 9, 7 and 1 of *B. cinerea* infection could be predicted from the simple measure $Z_{rel}$ of CW. The high $R^2_{McFadden}$ indicates that the impedance of CW is usable for the estimation of susceptibility of grapevines to *B. cinerea*. For further studies, it is an advantage that the impedance $Z_{rel}$ of CW is the most powerful phenotype, whereas the laborious removal of epicuticular waxes is not required. This is very important for common breeding questions because it enables the screening of hundreds of interesting genotypes in a short space of time by applying the simple to handle I-sensor.

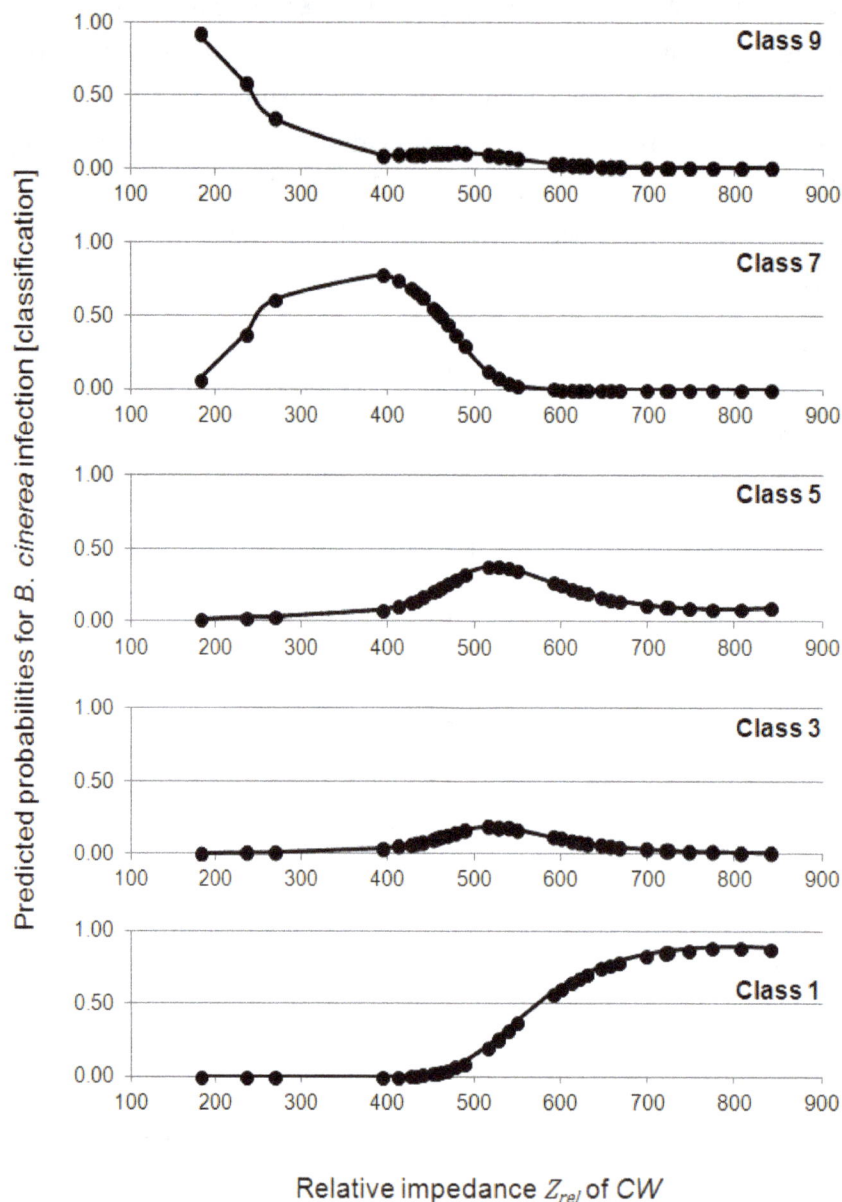

**Figure 7.** Prediction the probability of *B. cinerea* infection. Relative impedance $Z_{rel}$ of CW was applied in an ordinal logistic regression model. The data set including all genotypes (except genotypes with loose bunch compactness). Maximum Likelihood estimation was used and $R^2_{McFadden} = 0.37$ was calculated. CW: intact cuticle with epicuticular waxes.

For the generation of an improved regression model the set of investigated plants and further parameters should be included in the model, which may influence the susceptibility of grapevines to *B. cinerea*, e.g., the time of ripening or weather data (rainfall and temperature). In order to increase the objectivity of the model in the future, the *B. cinerea* infection of grapevine bunches could be determined with much more accuracy, e.g., by the application of objective, image-based phenotyping methods (an example is shown in Figure 8).

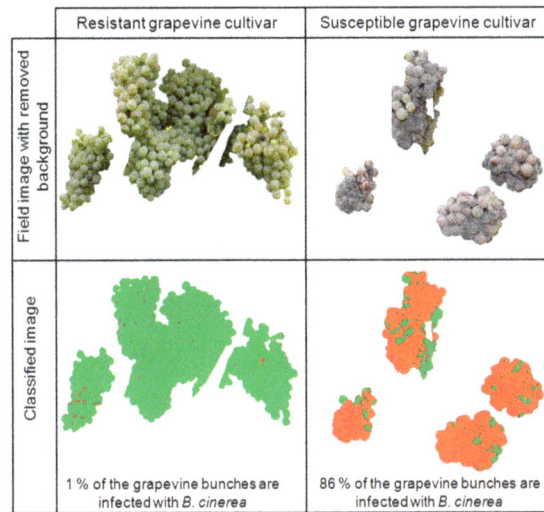

**Figure 8.** Exemplary image-based determination of *B. cinerea* infection on grapevine cultivars to improve reference evaluations of bunch rot in the field. The background of the field image was manually removed. Image segmentation into two phenotypic classes "healthy" (green) and "disease" (red) was performed by using Matlab®. The percentage amount of *B. cinerea* infection is quoted in the classified image.

*3.3. Impedance of the Berry Cuticle for QTL Analysis Applications*

The development of genetic markers for grapevine breeding purposes requires the acquisition of phenotypic traits from mapping populations (=F1 progeny) and the computation of significant QTL regions with a high LOD (logarithm of the odds) value. In the present study, the novel phenotypic trait of grape berry, the impedance of the cuticle, was tested for its application in QTL studies. Therefore, the phenotypic data ($Z_{rel}$ of C, W, CW and CCW) received from the F1 progeny was analyzed. Preliminary QTLs were identified whose LOD value were above the significance level (*i.e.*, LOD threshold). One preliminary QTL on Linkage Group (LG) 17 is illustrated in Figure 9.

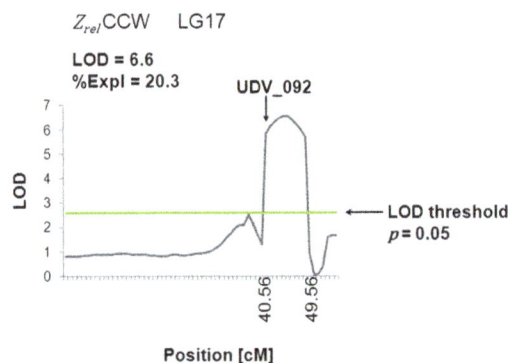

**Figure 9.** Preliminary QTL of the novel phenotypic trait in a F1 progeny (crossing of GF.GA.47-42 x 'Villard Blanc'). Genetic marker located within the detected QTL region is labeled with an arrow. The numbers on the x-axis indicate the position and confidence interval of the QTL region. LG = Linkage group; LOD = logarithm of the odds; %Expl = percentage of explained phenotypic variation. CCW: C (cuticle without epicuticular waxes) + CW (intact cuticle with epicuticular waxes).

For further investigations including the development of genetic markers for grapevine breeding applications, repeated phenotpying of this population is required to narrow down the QTL regions and to normalize the phenotypic data regarding environmental influences. In contrast to laborious, traditional phenotyping methods [11], the sensor-based technique developed in this study will enable precise phenotyping of a large mapping population with increased throughput. With the background that the impedance seems to be a promising indicator for bunch rot resistance, genetic markers could be established facilitating the selection of grapevine breeding material with a high impedance of the cuticle and epicuticular waxes implicating a genetic improvement of the resistance of breeding material to bunch rot.

## 4. Conclusions

Until now, resistance donors of genetic loci to increase the resistance of grapevines to *B. cinerea* are widely unknown. In common grapevine-breeding programs, the selection of seedlings with convenient physical properties (loose bunch architecture and small berries) can be done to increase the resistance to *B. cinerea*. In the present study, the simple-to-handle I-sensor was established and related to that, a fast phenotyping method was developed to estimate the resistance of grape bunches to *B. cinerea*. The I-sensor enables the measurement of the impedance of the cuticle and its epicuticular waxes from grape berries implementing a novel phenotypic trait. The identification of a preliminary QTL within a F1 progeny of the crossing of GF.GA.47-42 x 'Villard Blanc' shows the feasibility of this novel trait for genetic analysis. However, correlation studies show a high negative correlation between the impedance of cuticle and epicuticular waxes and the susceptibility of berries to *B. cinerea*. The novel sensor-based phenotyping technique developed in this study can thus be used to characterize the cuticles of grape berries in an easy and fast manner. The level of impedance indicates the susceptibility of a grapevine cultivar to *B. cinerea* and thus provides a novel important determinant with regard to bunch rot resistance. This effect was observed for genotypes with medium and compact bunch architecture. For future work, an increased number of genotypes and further parameter should be considered to improve the regression model, e.g., the consideration of the time of ripening or weather data. The phenotyping of the mapping population by applying the I-sensor should also be an important task for future work in order to determine new genetic markers for marker-assisted selection (MAS) of grapevine seedlings with improved resistance to bunch rot. In addition, the I-sensor could be applied in common breeding programs or other viticultural experiments, e.g., in order to evaluate the effect of canopy reduction or other vineyard management strategies on the berry skin quality.

## Acknowledgments

The work was supported by the AgroClustEr: CROP.SENSe.net funded by the German Federal Ministry of Education and Research (BMBF) within the scope of the competitive grants program Networks of excellence in agricultural and nutrition research (FKZ 0315534). The authors gratefully acknowledge Jan Behmann (Bonn University, Institute for Geoinformation and Geodesy) who had supported statistical discussions. Further, the authors thank Iris Ochßner and Hiltrud Heupel (Julius Kühn Institute) for grammatical editing the manuscript.

## Author Contributions

KH and RW created the measurement protocol. KH acquired and interpreted the phenotypic and reference data. KH also carried out statistical calculation and drafted the concept, text and figures of the manuscript. RW developed the I-sensor, corresponding software and the measurement principle. RW promoted the work for the present study with elementary ideas and experiments. RT initiated the project, revised the manuscript critically and was involved in data interpretation and discussion.

## Conflicts of Interest

The authors declare no conflict of interest.

## References

1. Jarvis, W.R. *Botryotinia and Botrytis Species: Taxonomy, Physiology, and Pathogenicity*; Canadian Department of Agriculture: Ottawa, ON, Canada, 1977; p. 195.
2. Molitor, D.; Behr, M.; Fischer, S.; Hoffmann, L.; Evers, D. Timing of cluster-zone leaf removal and its impact on canopy morphology, cluster structure and bunch rot susceptibility of grapes. *J. Int. Sci. Vigne Vin.* **2011**, *45*, 149–159.
3. Broome, J.; English, J.; Marois, J.; Latorre, B.; Aviles, J. Development of an infection model for botrytis bunch rot of grapes based on wetness duration and temperature. *Phytopathology* **1995**, *85*, 97–102.
4. Ellison, P.; Ash, G.; McDonald, C. An expert system for the management of botrytis cinerea in australian vineyards. I. Development. *Agric. Syst.* **1998**, *56*, 185–207.
5. Hed, B.; Ngugi, H.K.; Travis, J.W. Relationship between cluster compactness and bunch rot in vignoles grapes. *Plant Dis.* **2009**, *93*, 1195–1201.
6. Vail, M.; Marois, J. Grape cluster architecture and the susceptibility of berries to botrytis cinerea. *Phytopathology* **1991**, *81*, 188–191.
7. Vail, M.; Wolpert, J.; Gubler, W.; Rademacher, M. Effect of cluster tightness on botrytis bunch rot in six chardonnay clones. *Plant Dis.* **1998**, *82*, 107–109.
8. Molitor, D.; Behr, M.; Hoffmann, L.; Evers, D. Impact of grape cluster division on cluster morphology and bunch rot epidemic. *Am. J. Enol. Vitic.* **2012**, *63*, 508–514.
9. Nair, N.G.; Allen, R.N. Infection of grape flowers and berries by botrytis cinerea as a function of time and temperature. *Mycol. Res.* **1993**, *97*, 1012–1014.
10. Deytieux-Belleau, C.; Geny, L.; Roudet, J.; Mayet, V.; Donèche, B.; Fermaud, M. Grape berry skin features related to ontogenic resistance to botrytis cinerea. *Eur. J. Plant Pathol.* **2009**, *125*, 551–563.
11. Gabler, F.M.; Smilanick, J.L.; Mansour, M.; Ramming, D.W.; Mackey, B.E. Correlations of morphological, anatomical, and chemical features of grape berries with resistance to botrytis cinerea. *Phytopathology* **2003**, *93*, 1263–1273.
12. Kretschmer, M.; Kassemeyer, H.H.; Hahn, M. Age-dependent grey mould susceptibility and tissue-specific defence gene activation of grapevine berry skins after infection by botrytis cinerea. *J. Phytopathol.* **2007**, *155*, 258–263.

13. Nanni, V.; Schumacher, J.; Giacomelli, L.; Brazzale, D.; Sbolci, L.; Moser, C.; Tudzynski, P.; Baraldi, E. Vvamp2, a grapevine flower-specific defensin capable of inhibiting botrytis cinerea growth: Insights into its mode of action. *Plant Pathol.* **2014**, *63*, 899–910.

14. Commenil, P.; Brunet, L.; Audran, J.-C. The development of the grape berry cuticle in relation to susceptibility to bunch rot disease. *J. Exp. Bot.* **1997**, *48*, 1599–1607.

15. Becker, T.; Knoche, M. Deposition, strain, and microcracking of the cuticle in developing 'riesling' grape berries. *Vitis* **2012**, *51*, 1–6.

16. Becker, T.; Knoche, M. Water induces microcracks in the grape berry cuticle. *Vitis* **2012**, *51*, 141–142.

17. Schreiber, L. Transport barriers made of cutin, suberin and associated waxes. *Trends Plant Sci.* **2010**, *15*, 546–553.

18. Domínguez, E.; Heredia-Guerrero, J.A.; Heredia, A. The biophysical design of plant cuticles: An overview. *New Phytol.* **2011**, *189*, 938–949.

19. Benavente, J.; Ramos-Barrado, J.R.; Heredia, A. A study of the electrical behaviour of isolated tomato cuticular membranes and cutin by impedance spectroscopy measurements. *Colloids Surf. A Physicochem. Eng. Asp.* **1998**, *140*, 333–338.

20. Schreiber, L. Effect of temperature on cuticular transpiration of isolated cuticular membranes and leaf discs. *J. Exp. Bot.* **2001**, *52*, 1893–1900.

21. Schreiber, L.; Skrabs, M.; Hartmann, K.; Diamantopoulos, P.; Simanova, E.; Santrucek, J. Effect of humidity on cuticular water permeability of isolated cuticular membranes and leaf disks. *Planta* **2001**, *214*, 274–282.

22. Zyprian, E.; Eibach, R.; Töpfer, R. Eine neue genetische karte der weinrebe aus der kreuzung ''GF.GA-47–42'' x ''villard blanc''. *Deutsches Weinbau Jahrbuch.* **2006**, *57*, 151–158.

23. Fechter, I.; Hausmann, L.; Zyprian, E.; Daum, M.; Holtgräwe, D.; Weisshaar, B.; Töpfer, R. Qtl analysis of flowering time and ripening traits suggests an impact of a genomic region on linkage group 1 in vitis. *Theor. Appl. Genet.* **2014**, *127*, 1857–1872.

24. Zyprian, E.; Ochßner, I.; Schwander, F.; Simon, S.; Bonow-Rex, M.; Moreno-Sanz, P.; Grando, M.S.; Wiedemann-Merdinoglu, S.; Merdinoglu, D.; Eibach, R.; *et al.* Quantitative trait loci affecting resistance traits and ripening of grapevines in a genetic map based on single nucleotide polymorphisms and microsatellites. manuscript in preparation.

25. Lorenz, D.H.; Eichhorn, K.W.; Bleiholder, H.; Klose, R.; Meier, U.; Weber, E. Growth stages of the grapevine: Phenological growth stages of the grapevine (vitis vinifera l. Ssp. Vinifera)-codes and descriptions according to the extended bbch scale. *Aust. J. Grape Wine Res.* **1995**, *1*, 100–103.

26. OIV. OIV publications: OIV descriptor list for grape varieties and Vitis species (2nd ed.). Available online: http://www.Oiv.int (accessed on 26 May 2015).

27. Bühl, A. *Spss 16: Einführung in die Moderne Datenanalyse*; Pearson Deutschland GmbH: Hallbergmoos, Germany, 2008; Volume 7332.

# 6

# Reducing Systematic Centroid Errors Induced by Fiber Optic Faceplates in Intensified High-Accuracy Star Trackers

**Kun Xiong** [†] **and Jie Jiang** [†,*]

Key Laboratory of Precision Opto-Mechatronics Technology, Ministry of Education,
School of Instrumentation Science and Opto-Electronics Engineering, Beijing University of
Aeronautics and Astronautics (BUAA), Beijing 100191, China; E-Mail: xiongkun8748@163.com

[†] These authors contributed equally to this work.

[*] Author to whom correspondence should be addressed; E-Mail: jiangjie@buaa.edu.cn

Academic Editor: Anton de Ruiter

**Abstract:** Compared with traditional star trackers, intensified high-accuracy star trackers equipped with an image intensifier exhibit overwhelmingly superior dynamic performance. However, the multiple-fiber-optic faceplate structure in the image intensifier complicates the optoelectronic detecting system of star trackers and may cause considerable systematic centroid errors and poor attitude accuracy. All the sources of systematic centroid errors related to fiber optic faceplates (FOFPs) throughout the detection process of the optoelectronic system were analyzed. Based on the general expression of the systematic centroid error deduced in the frequency domain and the FOFP modulation transfer function, an accurate expression that described the systematic centroid error of FOFPs was obtained. Furthermore, reduction of the systematic error between the optical lens and the input FOFP of the intensifier, the one among multiple FOFPs and the one between the output FOFP of the intensifier and the imaging chip of the detecting system were discussed. Two important parametric constraints were acquired from the analysis. The correctness of the analysis on the optoelectronic detecting system was demonstrated through simulation and experiment.

**Keywords:** intensifier; fiber optic faceplate; intensified star tracker; systematic centroid error

# 1. Introduction

Star trackers are widely used in the attitude measuring systems of stationary satellites because of their advantages, which include high accuracy, non-drifting feature and automation [1–3]. Besides stationary applications, strategic missiles, high-maneuver satellites, *etc.* also employ star trackers [4,5] to provide high accuracy attitude information for gaining control. However, the star spots in an image will be dragged into the strips during maneuvering, or in worse cases, the strips will be so weak that they will be submerged in background noise.

Consequently, numerous discussions and studies [6,7] have been conducted on parameter optimization in optoelectronic detection systems to enhance their dynamic performance. Some researchers have claimed that the impact of maneuvering can be compensated by setting a relevant short period of exposure time followed by processing [8], while others employed optical lenses with large field of view (FOV) to observe more stars. Either short exposure or large FOV can improve the dynamic performance in some degree by means of tradeoffs. However, both tactics would cause decreasing sensitivity of optoelectronic systems as side effects. As a result, stars may not be observed by the star trackers employing these tactics when the maneuvering angular speed is high (5° per second or more). Revolutionarily, low light-level detectors with astounding increases in sensitivity were introduced as new-generation imaging devices for star trackers. Among such detectors, the intensified complementary metal oxide semiconductor (ICMOS) and the intensified charge coupled device (ICCD) exhibit the advantages of high sensitivity, small volume and low power consumption. In addition, these detectors have no requirement for extra cooling. Such advantages make ICMOS and ICCD ideal as imaging devices for the optoelectronic systems of star trackers. The intensifier consists of multiple FOFPs, hence, the optoelectronic detecting system is complicated and may cause accuracy loss in attitude measurement.

To date, a considerable number of analyses [9–13] have been performed on the influence of the intensifier on imaging. Given the traditional applications of the intensifier in night vision, the modulation transfer function (MTF), which describes the capability of a device to transfer different spatial frequency signals, is regarded as the key characteristic of optoelectronic systems. However, the performance of high-accuracy star trackers relies mainly on the accurate observation of starlight vectors derived from the centroid positions of star spots [14]. Compensation of systematic centroid errors in normal star trackers carried out in the frequency domain achieves sound results [15], while few studies on systematic centroid errors in the optoelectronic detecting system of intensified star trackers have been conducted. In Section 2, the imaging principle of the optoelectronic detecting system is described, and error sources related to FOFP are analyzed. In Section 3, a general expression of the systematic centroid error from an arbitrary optoelectronic component is obtained through centroid localization analysis in the frequency domain. Given that FOFP functions as the basic component of the intensifier, its image transmission model is presented in Section 4. Based on the analysis of FOFP, three FOFP-induced systematic error sources in the optoelectronic system for intensified star trackers are discussed in Section 5 to minimize possible errors that they could cause. Two parametric constraints are deduced from the discussion. The simulations and experiments presented in Section 6 validate the analysis made in this study.

## 2. Optoelectronic Detecting System of Intensified High-Accuracy Star Trackers

The imaging principle of the optoelectronic system is described in this section, particularly the low light-level enhancement mechanism in the intensifier. Different error sources related to FOFPs are analyzed, and FOFP-induced systematic centroid error is classified into three types.

### 2.1. Imaging Principle

The optoelectronic system of intensified high-accuracy star trackers works as follows: approximately infinitely far starlight enters the optical lens as parallel light. Passing through the lenses, the starlight converges into a Gaussian spot. Weak star spots enter the intensifier, and the intensity of the spots is enhanced. Complementary metal oxide semiconductor (CMOS) or charge coupled device (CCD) imaging chips coupled at the rear of the intensifier convert analog image signals into discrete digital image signals, which are transferred into the information processing system of the star tracker.

The core component of the detecting system is the intensifier. The intensified star tracker employed in this study uses the Generation II+ double-proximity focusing image intensifier, and its internal structure is shown in Figure 1. The photocathode supplies primary photoelectrons to the input face of the micro-channel plate (MCP) through its chemical coating, which transforms optical signals into electric signals. A few photoelectrons enter the MCP and are multiplied under the effect of the high-potential electric field. The secondary electrons from the MCP strike a phosphor screen that emits photons.

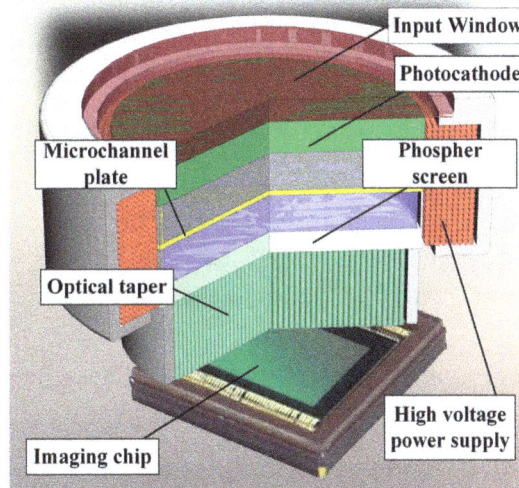

**Figure 1.** Internal structure of the intensifier [16].

### 2.2. FOFP-Induced Centroid Error

Both the photocathode and the anode of the intensifier normally bond to an FOFP. Similar to an FOFP, the MCP also arranges its micro-channels hexagonally, and thus, it is regarded as the equivalent of FOFP for electrons. In addition, the photons emitting from the screen are optically coupled to the imaging chips by an FOFP, which is called an optical taper. That is, it is the intensifier that is a compound with multiple FOFPs. For intensified star trackers, addition of an image intensifier will cause complex centroid errors. Firstly, the gain variation attributed to the quantum effect in the electron multiplying process of MCP causes centroid errors that change over time. Secondly, the defection of fiber arrangement in FOFP causes centroid errors that change with position. These two error sources cannot be described by specific expressions, and thus, they are difficult to correct or reduce. Thirdly, given that FOFP is actually a discrete optical component that is neither isoplanatic nor shift invariant, it can cause systematic centroid errors that are distinct at different places.

In an optoelectronic detecting system, the intensifier is located between the optical lens and the imaging chip, as shown in Figure 2. Three error sources are found along the entire optical path when the systematic centroid errors of an FOFP are involved: Error 1: Systematic errors occur when the star spot from the optical lens enters the input FOFP of the intensifier. Error 2: The multiple FOFPs of the intensifier will lead to even more complex systematic errors. Error 3: The star spot that is emitting from the output FOFP of the intensifier enters imaging chip coupling from behind and generates systematic errors because of the sampling of the imaging chip [17]. The errors caused by the aforementioned sources are collectively referred to as FOFP-induced systematic centroid errors in this study. All these errors are covered by the following investigation.

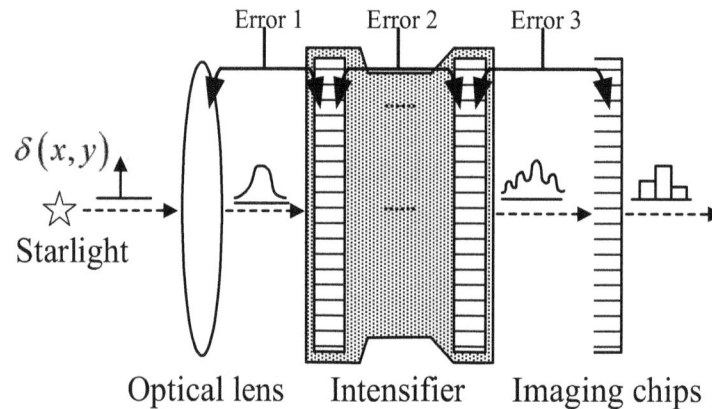

**Figure 2.** Diagram of the FOFP-induced systematic centroid errors.

## 3. Frequency Domain Analysis of Centroid Localization

The direct error analysis of a sophisticated optoelectronic system can be complicated, and thus, frequency domain analysis is a better alternative. The expressions of centroid localization and the centroid error of an arbitrary optical component are presented in this section and function as the basis for the following discussion on FOFP-induced systematic centroid error.

### 3.1. Centroid Localization Expression in the Frequency Domain

The subdivided centroid localization methods for star trackers can be generally divided into the mass center method and the surface fitting method. To improve star spot position estimation, several new methods consider the weight functions or change the order of the moment in the mass center method, whereas others change the surface expression in the surface fitting method. When real time is considered, the mass center method has been widely used because of its advantages of small computational cost and easily realized algorithm structure. The discussion on centroid error in this paper is specific to the mass center method.

The mass center method in centroid localization assumes that the exact location of a star spot coincides with its mass center of luminance. Given that an image is discretized into pixels by the imaging sensor, the mass center method is always expressed in summation form. More generally in the continuous spatial domain, by denoting the intensity distribution of a star spot as $f(x,y)$, the centroid location of the spot is given as follows:

$$\overline{x} = \frac{\iint xf(x,y)\,dxdy}{\iint f(x,y)\,dxdy} \quad \overline{y} = \frac{\iint yf(x,y)\,dxdy}{\iint f(x,y)\,dxdy} \tag{1}$$

where $x$ and $y$ are the metrics of two orthogonal spatial axes. By applying Fourier transform to the distribution function $f(x,y)$, its frequency domain expression is given as follows:

$$F(u,v) = \int_{-\infty}^{+\infty}\int_{-\infty}^{+\infty} f(x,y)\exp\left[-j2\pi(ux+vy)\right]dxdy \tag{2}$$

where $u$ and $v$ are the frequency metrics that correspond to two orthogonal spatial axes. Deducing that the partial derivative of $F(u,v)$ with respect to $u$ is expressed as follows:

$$F_u(u,v) = \frac{\partial F}{\partial u} = -j2\pi \cdot \int_{-\infty}^{+\infty}\int_{-\infty}^{+\infty} xf(x,y)\exp\left[-j2\pi(ux+vy)\right]dxdy \tag{3}$$

In the origin of the frequency plane, $F(u,v)$ and its partial derivative $F_u(u,v)$ can be simplified as follows:

$$F(0,0) = \int_{-\infty}^{+\infty}\int_{-\infty}^{+\infty} f(x,y)\exp\left[-j2\pi(0x+0y)\right]dxdy = \int_{-\infty}^{+\infty}\int_{-\infty}^{+\infty} f(x,y)dxdy \tag{4}$$

$$F_u(0,0) = -j2\pi \cdot \int_{-\infty}^{+\infty}\int_{-\infty}^{+\infty} xf(x,y)dxdy \tag{5}$$

By comparing the two preceding expressions with the expression of the mass center method in the spatial domain (shown in Equation (1)), we obtain:

$$\frac{F_u(0,0)}{F(0,0)} = \frac{-j2\pi \cdot \int_{-\infty}^{+\infty}\int_{-\infty}^{+\infty} xf(x,y)dxdy}{\int_{-\infty}^{+\infty}\int_{-\infty}^{+\infty} f(x,y)dxdy} = -j2\pi\overline{x} \tag{6}$$

The expression is simplified as follows:

$$\overline{x} = \frac{F_u(0,0)}{-j2\pi F(0,0)} \tag{7}$$

Equation (7) associates the centroid location with the spectrum expression of the spot in the frequency domain. According to the equation, the centroid location is only related to the origin point values of the star spot spectrum in the frequency domain and its partial derivative.

## 3.2. Systematic Centroid Error Expression

Assuming that the spot pattern goes through a shift-invariant optoelectronic component, the Point Scattering Function (PSF) of the part is denoted as $h(x,y)$, and the distribution of the output image $g(x,y)$ is given as follows:

$$g(x,y) = f(x,y) * h(x,y) \tag{8}$$

In the frequency domain, the course can be expressed as a direct multiplication instead of a convolution as follows:

$$G(u,v) = F(u,v) \cdot H(u,v) \tag{9}$$

From Equation (7), the centroid location that is going through the preceding system will change as follows:

$$\tilde{x} = \frac{G_u(0,0)}{-j2\pi G(0,0)} = \frac{F_u(0,0)H(0,0)+F(0,0)H_u(0,0)}{-j2\pi F(0,0)H(0,0)} = \bar{x} + \frac{H_u(0,0)}{-j2\pi H(0,0)} \tag{10}$$

Thus, the systematic centroid of the component is as follows:

$$\Delta x = \tilde{x} - \bar{x} = \frac{H_u(0,0)}{-j2\pi H(0,0)} \tag{11}$$

According to Equation (11), when starlight comes through an optoelectronic component, the involved centroid error only depends on the system itself. Furthermore, for an isoplanatic component (which is also translation invariant), it is easy to infer that the systematic error it brings about is constant in the image plane.

However, in a shift-variant system, PSF varies from place to place throughout the image plane. PSF is denoted as $h\big|_{(\bar{x},\bar{y})}(x,y)$, where $(\bar{x},\bar{y})$ represents the location of the input spot $f(x,y)$. Meanwhile, its expression in the frequency domain is denoted as $H\big|_{(\bar{x},\bar{y})}(u,v)$. Following Equation (11), the systematic error of the shift-variant system is given as follows:

$$\Delta x_{(x_c,y_c)} = \frac{H_u\big|_{(x_c,y_c)}(0,0)}{-j2\pi H\big|_{(x_c,y_c)}(0,0)} \tag{12}$$

According to the preceding equation, the systematic centroid error in the shift-variant system varies with the position of the input object. In particular, the discrete optic component has microcells that repeat periodically in certain spatial structures, and thus, the systematic error that it causes is also periodic.

## 4. Image Transmission Model of FOFP

FOFP is a common image-carrying component that consists of a huge amount of fibers arranged in a hexagonal grid structure similar to a honeycomb. It is optically equivalent to a zero-thickness window. As mentioned in Section 2, an intensifier mainly consists of multiple FOFPs; hence, analyzing FOFP is highly important. A diagram that depicts FOFP structure is shown in Figure 3.

A cutaway diagram of a single fiber is shown in Figure 4. The incident photons that are hitting the end face of the shield wrap will not get through (shown in dotted arrows), whereas those entering the fiber core will pass through via total reflection (shown in solid arrows). The fibers are carefully arranged to ensure that image transmission will not have any distortion. A lightproof shield wrap that surrounds the fiber core is used to block cross disturbance among fibers. In addition, fibers are typically cylindrical; therefore, small gaps exist between neighboring fibers because of the hexagonal arrangement of the structure. Neither the shield wrap nor the small gaps can transmit light.

Consequently, these dead zones in the optic fiber plate will affect both the imaging and the centroid localization of high-accuracy star trackers.

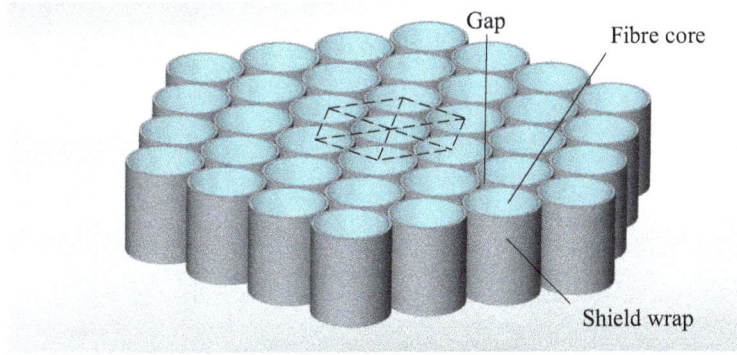

**Figure 3.** Diagram of FOFP structure.

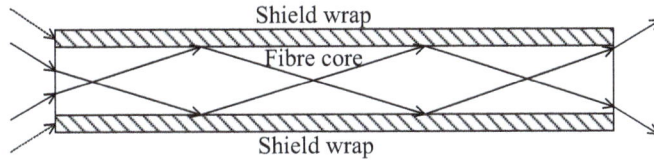

**Figure 4.** Cutaway diagram of a single fiber.

Disregarding its thickness, absorption loss, transmission spectrum *etc.*, an FOFP can be regarded as a mask of round holes with a particular arrangement. When an image is passing through an FOFP, it follows the processes of integration, sampling and reimaging in order. In the first two steps, the light entering an FOFP converges its intensity in the fiber cores and forms energetic impulses. In reimaging, the energy pulse emerges from the fiber core in a certain pattern. The PSF of FOFP $h(x,y)$ can be expressed as follows:

$$h(x,y) = h_{samp}(x,y) * h_{re}(x,y) \tag{13}$$

The input end face of a single fiber is a circle, and thus, the integration process can be described by cylinder function circ($r$) as follows:

$$\text{circ}(r) = \text{circ}(\sqrt{x^2 + y^2}) = \begin{cases} 1 & r \leq 1 \\ 0 & r > 1 \end{cases} \tag{14}$$

A diagram of the 'mask plate' that represents the FOFP mechanism is provided in Figure 5, where the radius of the fiber core is denoted as $r_0$ and the fiber interval is $a$. The arrangement of the fibers is shown in the figure, and the size of the FOFP is a square of $W \times W$. Then, the integration–sampling function of the FOFP can be expressed as follows:

$$h_{samp} = \left[ \text{circ}\left(\frac{r}{r_0}\right) \cdot \text{comb}\left(\frac{x}{\sqrt{3}a} + \frac{y}{a}, \frac{x}{\sqrt{3}a} - \frac{y}{a}\right) \right] \cdot \text{rect}\left(\frac{x}{W}, \frac{y}{W}\right) \tag{15}$$

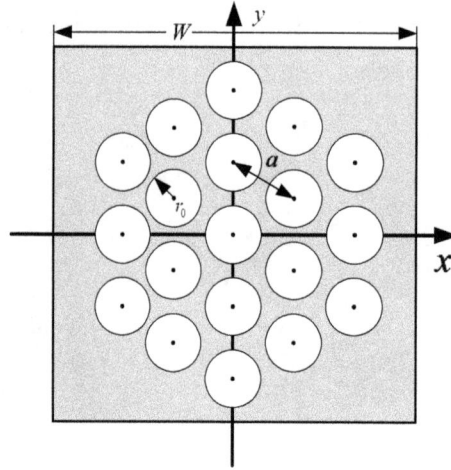

**Figure 5.** FOFP model.

The reimaging process is complicated. Reimaging varies with different optical paths inside the fiber caused by various incident angles and incident positions of the incident light. Under ideal circumstances, the energy is believed to output fibers uniformly. Hence, the reimaging function can be written as follows:

$$h_{re} = \text{circ}\left(\frac{r}{r_0}\right) \tag{16}$$

According to Equations (13)–(16), the PSF of an FOFP can be expressed as follows:

$$h(x,y) = \left[\text{circ}\left(\frac{r}{r_0}\right) \cdot \text{comb}\left(\frac{x}{\sqrt{3}a} + \frac{y}{a}, \frac{x}{\sqrt{3}a} - \frac{y}{a}\right) \cdot \text{rect}\left(\frac{x}{W}, \frac{y}{W}\right)\right] * \text{circ}\left(\frac{r}{r_0}\right) \tag{17}$$

where $r = \sqrt{x^2 + y^2}$.

The Fourier transform of the PSF is the modulation transfer function (MTF), *i.e.,*:

$$H(u,v) = \left[\frac{J_1(2\pi r_0 f)}{\pi r_0 f} * \text{comb}\left(\frac{\sqrt{3}a}{2}u + \frac{a}{2}v, \frac{\sqrt{3}a}{2}u - \frac{a}{2}v\right) * \text{sinc}(Wu, Wv)\right] \cdot \frac{J_1(2\pi r_0 f)}{\pi r_0 f} \tag{18}$$

where $f = \sqrt{u^2 + v^2}$. Given that the size of an FOFP is significantly greater than the scale of a fiber, the frequency transform of the rect function, *i.e.*, sinc($Wu,Wv$), can be considered as a unit of impulse function, as follows:

$$\lim_{W \to \infty} \text{sinc}(Wu, Wv) = \delta(u,v) \tag{19}$$

Then, the MTF of an FOFP becomes:

$$H(u,v) = \left[\frac{J_1(2\pi r_0 f)}{\pi r_0 f} * \text{comb}\left(\frac{\sqrt{3}a}{2}u + \frac{a}{2}v, \frac{\sqrt{3}a}{2}u - \frac{a}{2}v\right)\right] \cdot \frac{J_1(2\pi r_0 f)}{\pi r_0 f}. \tag{20}$$

## 5. Reducing FOFP-Induced Systematic Centroid Errors

As discussed in Section 2.2, three types of FOFP-induced systematic centroid errors occur in the optoelectronic detecting system of intensified star trackers. Basically, each error is determined by the match between a component and the succeeding one. The matches are discussed by means of a frequency spectrum. The suppression of all the aforementioned error sources is analyzed chronologically in this section.

### 5.1. Systematic Error Reduction between the Optical Lens and the Input FOFP of an Intensifier

#### 5.1.1. Optical Lens Analysis

Theoretically, starlight distribution before the optical lens can be regarded as an energetic impulse function $E\delta(x-x_0, y-y_0)$. When the lenses are modulated, the impulse point blurs into a spot. Assuming that the lenses are equivalent to an ideal convex lens, and the PSF of the lens $h_0(x, y)$ will be the same Gaussian function everywhere, as follows:

$$h_0(x,y) = \frac{1}{2\pi\sigma^2} e^{\frac{-(x^2+y^2)}{2\sigma^2}}$$

(21)

where $3\sigma$ is the radius of the spot. The MTF of the lenses is as follows:

$$H_0(u,v) = e^{-2\pi^2\sigma^2(u^2+v^2)}$$

(22)

Then, the star spot behind the lenses can be written as follows:

$$f(x,y) = \left[E\delta(x-x_0, y-y_0)\right] * h_0(x,y) = \frac{E}{2\pi\sigma^2} e^{-\frac{(x-x_0)^2+(y-y_0)^2}{2\sigma^2}}$$

(23)

The ideal optical lens is translation invariant; hence, it has no effect on centroid localization according to Equation (10), as follows:

$$\Delta x = \frac{F_u(0,0)}{-j2\pi F(0,0)} = 0, \ \Delta y = \frac{F_v(0,0)}{-j2\pi F(0,0)} = 0$$

(24)

That is, centroid localization will remain error-free after passing through the lenses.

#### 5.1.2. Systematic Centroid Error of an FOFP

According to the previous analysis, the effect of an FOFP on the input image can be divided into three separate processes: integration, sampling and reimaging. The reimaging process is a translation-invariant cylinder function (as shown in Equation (16)). According to the analysis in Section 3.2, no systematic error will be evolved by the reimaging process. That is, the systematic centroid error caused by an FOFP is formed during integration and sampling.

In the integration and sampling processes, the input image initially convolves with the cylinder function circ($r$), and then, the result is sampled by a hexagonal comb function. The transform of the spatial hexagonal comb function in the frequency domain is still a hexagonal comb function. Any image pattern that goes through the plate will have its frequency spectrum replicated and overlapped.

The original frequency spectrum will be replicated to areas centered on each frequency point shown in the shaded area in Figure 6. These frequency points are assigned to a hexagonal grid structure that is perpendicular to one of the sample points in the spatial domain, with a distance of $2/\sqrt{3}a$ (approximately $1.15/a$) between two neighboring points. These points are called overlapping frequency points.

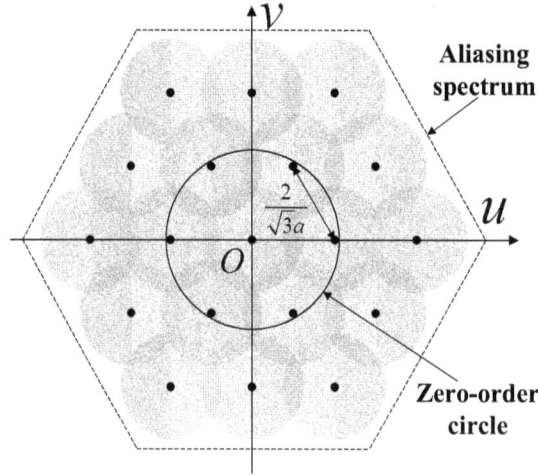

**Figure 6.** Overlapping frequency points in the frequency domain.

The overlapping of the frequency spectrum can cause frequency aliasing. The effect of frequency aliasing on centroid localization is analyzed in this section. Assuming that starlight spot input in the FOFP obeys Gaussian distribution, the exact spot in the origin spot can be denoted as $f_0(x, y)$, and then, the distribution $f(x, y)$ of a starlight spot in the arbitrary position $(\overline{x}, \overline{y})$ can be expressed as follows:

$$f(x,y) = f_0(x - \overline{x}, y - \overline{y}) \tag{25}$$

After integration, the PSF is convolved with the cylinder function as follows:

$$g(x,y) = f(x,y) * \mathrm{circ}\left(\frac{r}{r_0}\right) = f_0(x - \overline{x}, y - \overline{y}) * \mathrm{circ}\left(\frac{r}{r_0}\right) \tag{26}$$

After being transformed in the frequency domain, the Gaussian spectrum is multiplied by the first-order Bessel curved surface as follows:

$$G(u,v) = F(u,v) \cdot \frac{J_1(2\pi r_0 f)}{\pi r_0 f} = e^{-j2\pi(u\overline{x} + v\overline{y})} \cdot G_0(u,v) \tag{27}$$

where $G_0(u, v)$ represents the after-integration frequency spectrum of the star spot at the origin point. It would still be rotationally symmetric to the origin point in the frequency domain. After being transformed into frequency polar coordinates, spectrum $G_0(u, v)$ and its partial derivative can be expressed as follows:

$$\begin{cases} G_0(f,\theta) = G_0(f) \\ G_{0u}(f,\theta) = \cos(\theta) G_0'(f) \\ G_{0v}(f,\theta) = \sin(\theta) G_0'(f) \end{cases} \tag{28}$$

As shown in Figure 6, assuming that only the first-order overlapping spectra will interfere with the origin spectrum, then the aliasing spectrum $\tilde{G}(u,v)$ after the sampling process can be written as follows:

$$\tilde{G}(u,v) = \sum_{n=-1}^{+1} G(u-nD,v) + \sum_{m=0}^{+1}\sum_{k=0}^{+1} G\left(u-mD-\frac{D}{2}, v-\sqrt{3}kD-\frac{\sqrt{3}D}{2}\right) \tag{29}$$

where $D$ represents the intervals between two overlapping frequency points (*i.e.*, $2/\sqrt{3}a$. According to Equations (28)–(29), we obtain

$$\tilde{G}(0,0) = G_0(0) + 2\left[\cos(2\pi D\overline{x}) + 2\cos(\pi D\overline{x})\cos(\sqrt{3}\pi D\overline{y})\right]G_0(D) \tag{30}$$

$$\tilde{G}_u(0,0) = -j2\pi\overline{x}G(0,0) + 2jG_0'(D)\left[\sin(2\pi D\overline{x}) + \sin(\pi D\overline{x})\cos(\sqrt{3}\pi D\overline{y})\right] \tag{31}$$

$$\tilde{G}_v(0,0) = -j2\pi\overline{y}G(0,0) + 2\sqrt{3}jG_0'(D)\cos(\pi D\overline{x})\sin(\sqrt{3}\pi D\overline{y}) \tag{32}$$

Compared with the first item in $\tilde{G}(0,0)$, the second one is significantly smaller. From Equation (12), the estimated centroid location under the aliasing influence of the first-order overlapping spectra can be approximately expressed as follows:

$$\tilde{x} \approx \overline{x} - \frac{G_0'(D)\left[\sin(2\pi D\overline{x}) + \sin(\pi D\overline{x})\cos(\sqrt{3}\pi D\overline{y})\right]}{\pi G_0(0) + 2\pi\left[\cos(2\pi D\overline{x}) + 2\cos(\pi D\overline{x})\cos(\sqrt{3}\pi D\overline{y})\right]G_0(D)} \tag{33}$$

$$\tilde{y} \approx \overline{y} - \frac{\sqrt{3}G_0'(D)\cos(\pi D\overline{x})\sin(\sqrt{3}\pi D\overline{y})}{\pi G_0(0) + 2\pi\left[\cos(2\pi D\overline{x}) + 2\cos(\pi D\overline{x})\cos(\sqrt{3}\pi D\overline{y})\right]G_0(D)} \tag{34}$$

The magnitude in frequency $D$ is considerably less than that in the origin, and thus, the two preceding equations can be approximated as follows:

$$\tilde{x} = \overline{x} - \frac{G_0'(D)}{\pi G_0(0)}\left[\sin(2\pi D\overline{x}) + \sin(\pi D\overline{x})\cos(\sqrt{3}\pi D\overline{y})\right] \tag{35}$$

$$\tilde{y} = \overline{y} - \frac{G_0'(D)}{\pi G_0(0)}\sqrt{3}\cos(\pi D\overline{x})\sin(\sqrt{3}\pi D\overline{y}) \tag{36}$$

The actual center of the input starlight spot is known as $(\overline{x}, \overline{y})$; hence, the systematic centroid error caused by the FOFP is as follows:

$$\Delta x(\overline{x}, \overline{y}) = -\frac{G_0'(D)}{\pi G_0(0)}\left[\sin(2\pi D\overline{x}) + \sin(\pi D\overline{x})\cos(\sqrt{3}\pi D\overline{y})\right] \tag{37}$$

$$\Delta y(\overline{x}, \overline{y}) = -\frac{G_0'(D)}{\pi G_0(0)}\sqrt{3}\cos(\pi D\overline{x})\sin(\sqrt{3}\pi D\overline{y}) \tag{38}$$

According to the result, the overall magnitude of the systematic error is determined by the ratio of the derivative value in the first-order overlapping points $G_0'(D)$ to the spectrum value in the origin point $G_0(0)$. The repetition period of the systematic error is determined by the distance $D$ between two neighboring overlapping frequency points. As shown in Figure 7, the systematic error magnitude distribution with starlight spots enters the FOFP from different places. If the spots are located at the center of a fiber or at the border of two neighboring fibers, then they would be affected by a minor systematic error. When the spots move from the border area and towards the fiber center, the error magnitude will initially increase and then decrease.

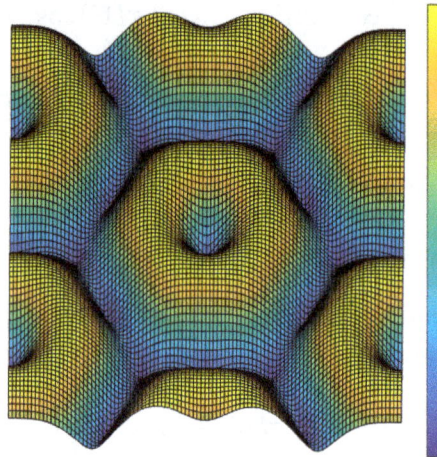

**Figure 7.** Magnitude of the systematic centroid error with respect to the entrance position of the starlight spot.

5.1.3. Error Reduction

The integration process causes the Gaussian spot to become convoluted with the cylinder function. The Fourier transform of the cylinder function is the first-order Bessel function. Nearly all of the energy of this function is generally acknowledged to focus within the zero-order central circular area with a diameter of $1.22/r_0$. The sampling process will cause the original frequency spectrum to shift and overlap, with the interval of two neighboring overlapping frequency points being $1.15/a$. Disregarding the thickness of the shield wrap ($r_0 = a$), the diameter of the spectrum after integration becomes:

$$1.22/r_0 = 1.22/a > 1.15/a \tag{39}$$

If the thickness of the shield wrap is considered, then the spectrum diameter will be larger. In the analysis of the frequency domain, the accurate recovery of an image requires no aliasing in the spectrum after sampling. This condition is the essence of Shannon's sampling theorem. However, the only requirement of a star tracker is that no systematic centroid error will result when an image passes through the component. According to the conclusion in Section 3.2, this condition indicates that the value at the origin point of the spectrum and its derivative value should not be affected by the component. As shown in Figure 6, the circle marks the minimum border of the zero-order central circle of the first-order Bessel function, within which lies six first-order overlapping shift points. Therefore, if no constraint is placed upon the star spot that is entering the input FOFP, then systematic error will be inevitable. The constraint to the spectrum of the entering image $F(u,v)$ is as follows:

$$F(u,v)\big|_{f \geq 1.15/a} = 0 \qquad (40)$$

The spectrum of the starlight spot from the lenses in the frequency domain still obeys Gaussian distribution. The star spot described by Equation (23) will have its frequency spectrum distribution within a circle that is $6\sigma$ in diameter. Thus, the upper frequency limit is as follows:

$$f_H = 3 \cdot \frac{1}{2\pi\sigma} \qquad (41)$$

By combining Equations (40) and (41), we obtain

$$6\sigma \geq 6 \cdot \frac{3}{2\pi} \cdot \frac{a}{1.15} \approx 2.49a \qquad (42)$$

According to Equation (42), the circle of confusion ($6\sigma$) should not be less than 2.49 times that of the fiber interval to free the star tracker from systematic errors caused by the input FOFP.

*5.2. Systematic Error Reduction among Multiple FOFPs*

In general, the intensifier contains more than one FOFP. This structure can cause more serious aliasing problems than that of a single FOFP. Take two FOFPs as an example. Assume that the fiber core radius $r_0$ and the fiber interval $a$ in all the FOFPs are the same. As shown in Figure 8, the black dots represent the frequency of overlapping shift points of the first FOFP. The black line-covered area describes the first-order frequency aliasing of the first FOFP. The black crosses represent the frequency of overlapping points of the second FOFP, whose fiber arrangement direction is assumed to be 15° tilting from the first FOFP. The gray shade represents the overlapping spectrum from one of the six first-order overlapping frequency points. The value in the origin point of the frequency domain becomes extremely complex after aliasing.

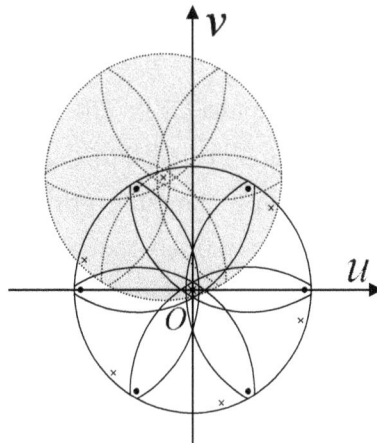

**Figure 8.** Frequency spectrum aliasing that causes centroid error.

To avoid the aforementioned serious frequency spectrum aliasing, the highest frequency of the image spectrum from the previous FOFP must be less than the overlapping frequency of the next

FOFP. An optical low-pass filter (OLPF) should be placed between two neighboring FOFPs. In practice, the reimaging process of an FOFP is more complicated than the description provided in Equation (16). Moreover, a certain amount of gap between two neighboring FOFPs is designed on purpose to defocus the image. This consideration somehow functions as an OLPF, and the frequency spectrum of the entering image can be limited within the overlapping frequency. The situation in which no systematic centroid error will be caused by multiple FOFPs in the intensifier is shown in Figure 9.

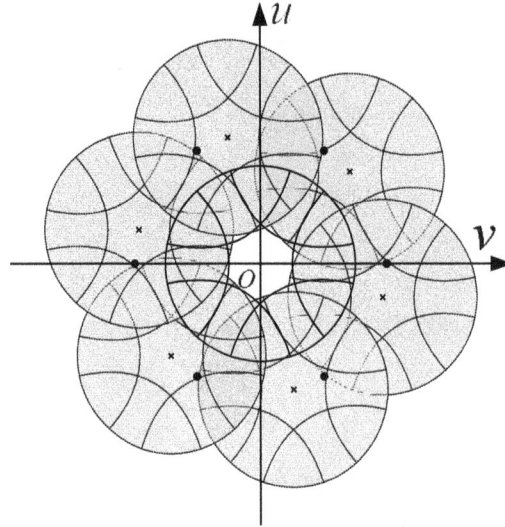

**Figure 9.** Frequency spectrum aliasing that is free from centroid error.

*5.3. Systematic Error Reduction between the Output FOFP of the Intensifier and the Imaging Chip*

The output FOFP is typically referred to as the optic taper coupling with the imaging chip. It transfers the enhanced image of the internal fluorescent screen to the CCD or CMOS chip behind it. Under the effect of the reimaging process (as shown in Equation (16)), the frequency spectrum of the image passing through the output FOFP lies within a circular area whose diameter $d_{freq}$ is:

$$d_{freq} = 0.61 / r_{0N} \tag{43}$$

where $r_{0N}$ is the fiber core diameter of the output FOFP. The image passing through the output FOFP will be sampled by the imaging chip and transformed into digital signals. The pixels of a CMOS or CCD imaging sensor are typically arranged in rectangular grids. Assuming that the sampling interval of the imaging chip is $d$, inferring that the overlapping frequency is $1/d$ is easy. Similar to the analysis presented in Section 5.1, if no systematic error is caused by the imaging chip, then the spectrum radius that is being outputted from the intensifier should not exceed the overlapping frequency, which is as follows:

$$\frac{1}{d} \geq d_{freq} \tag{44}$$

$$d \leq \frac{r_{0N}}{0.61} = 1.64 r_{0N} \tag{45}$$

When the pixel size of the coupling imaging chip satisfies the constraint in Equation (45), the sampling frequency is sufficient to ensure that no systematic centroid error will occur in the imaging chips.

## 6. Simulations and Experiments

Simulations and an experiment were conducted to verify the major conclusions drawn from the analysis of the reduction of FOFP-induced systematic centroid errors. Firstly, the simulation of a star spot entering at different places of an FOFP was presented, and the systematic centroid error distribution of the FOFP with respect to the entering position of the spot was compared with the expressions obtained in Section 5.1.2. Secondly, two simulations were performed to verify the constraint between the optical lens and the input FOFP of the intensifier. Thirdly, the calibration results of two intensified star tracker prototypes with different imaging chips were compared to demonstrate the analysis of systematic errors between the output FOFP of the intensifier and the imaging chip.

### 6.1. Simulation of the Systematic Centroid Error of an FOFP

A simulation of the systematic centroid error of an FOFP under different star spot positions was performed. The simulated parameter configuration of the star tracker is listed in Table 1. The simulation was conducted in a square FOFP area with 20 μm × 20 μm, and focused on a single fiber. The test star spots that were entering the FOFP were supposed to be Gaussian distributed, and their density was 0.2 μm in both directions. Following the FOFP model described in Section 4, integration and sampling were conducted on the original star spot and redistributed following the reimaging function.

**Table 1.** Simulation parameter configuration.

| Parameters | Values |
|---|---|
| Fiber core diameter | 5.5 μm |
| Fiber interval | 6 μm |
| Star spot diameter | 12 μm |
| Simulation area | 20 μm × 20 μm |
| Test star spot density | 0.2 μm |

The simulation result is shown in Figure 10. Figure 10a presents the error vector distribution of the systematic error. The circles in the figure show where the fiber cores are. If the center of a star spot lies within the fiber cores, then the direction of the centroid error vector is radial. If the center of a spot falls within the border area of neighboring fibers, then systematic error decreases because of the counteracting effect. Figure 10b shows the error magnitude distribution. The maximum error can reach over 1 μm, which is obviously non-ignorable in high-accuracy star trackers. Moreover, the distribution pattern is approximately the same as the one deduced in the frequency domain (shown in Equations (29) and (30) and in Figure 7). The small bumps on the curved surface are caused by the boundary effect of the simulation area.

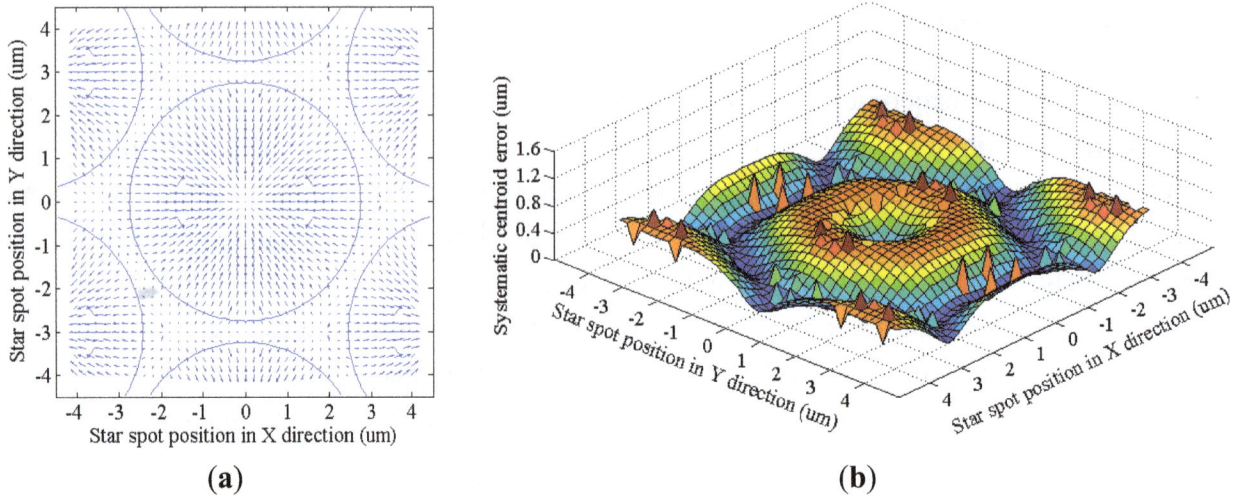

**Figure 10.** Systematic centroid error of the FOFP. (**a**) Error vector distribution; (**b**) Error magnitude distribution.

*6.2. Simulation of the Match between the Optical Lens and the Input FOFP of the Intensifier*

According to Equation (42), certain proportional relation between the circle of confusion of the lenses and the fiber interval of the FOFP must be satisfied. The following simulation kept all the values of the FOFP parameters unchanged while altering the circle of confusion. The systematic centroid error located at (0.8 μm, 1.4 μm) was observed through the changes. The chosen location is neither near the center of the fiber nor near the border area of fibers where the systematic centroid error is believed to be small, so it can represent the error magnitude better.

As shown in Figure 11, the systematic error decreases rapidly as the circle of confusion of the entering star spot grows. If the circle of confusion is larger than the critical condition deduced from Equation (42) (14.94 μm), then systematic error magnitude will not exceed 0.1 μm (shaded area in the figure). The equivalent angular error is typically significantly less than 1 arc-second, and thus, it can be ignored. The simulation result verified that the circle of confusion of the optical lens should be 2.49 times that of the input FOFP fiber interval or more.

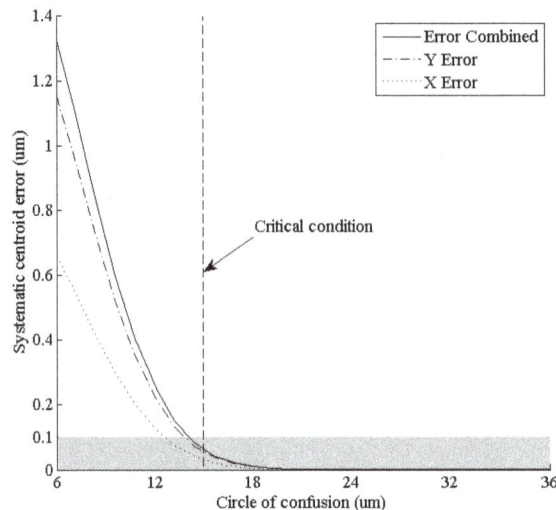

**Figure 11.** Systematic centroid error at (0.8 μm, 1.4 μm) under different circles of confusion.

Not only the fiber interval but also the core-sheath ratio would affect the systematic centroid error. A simulation under different fiber core radii $r_0$ was conducted, with all the other parameters unchanged in Sections 6.1. The systematic centroid error located at (0.8 μm, 1.4 μm) was observed through the changes.

As shown in Figure 12, the systematic error decreases as the fiber core radius grows. The decrease is nearly linear. When the fiber core radius reaches 3 μm (the upper limit caused by the fiber interval), the error magnitude is approximately half of the fiber core with a radius of 1 μm fiber core. That is, increasing the core-sheath ratio as much as possible under a fixed fiber interval will help harness systematic centroid error.

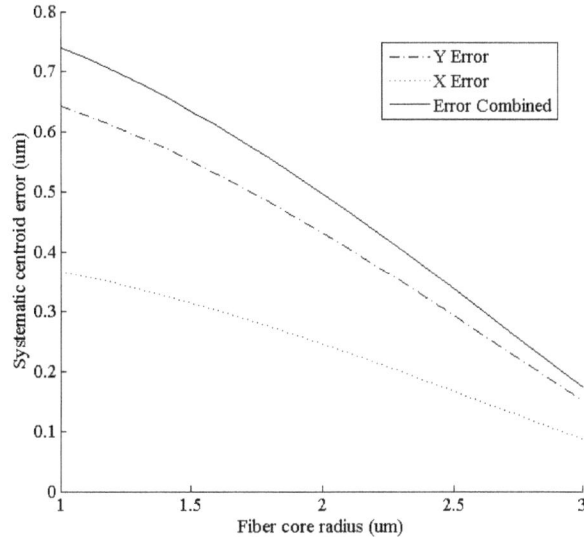

**Figure 12.** Systematic centroid error at (0.8 μm, 1.4 μm) under different fiber core radii.

*6.3. Experimental Investigation on the Match between the Output FOFP of the Intensifier and the Imaging Chip*

Two prototypes of intensified star trackers were compared in terms of systematic centroid error. The imaging chips of the two star trackers are different. The first adopted a CCD chip from DALSA Inc. (Waterloo, ON, Canada), whereas the second chose a CMOS chip from CMOSIS Co (Antwerp, Belgium). The details are provided in Table 2. The pixel size of the CCD is half that of the CMOS, whereas both image sizes are 1 optical inch. Both the image chips were coupled with a same 2nd-generation double proximity image intensifier that was 18 mm in diameter. The output FOFP of the intensifier has a fiber core diameter of 5.5 μm and a fiber interval of 6 μm.

**Table 2.** Comparison of the two imaging chips.

| Chip Name | FTT1010M | CMV4000 |
|---|---|---|
| Manufacturer | Dalsa Inc. | CMOSIS Co. |
| Chip Type | Frame Transfer CCD | CMOS |
| Pixel Size | 12 μm × 12 μm | 5.5 μm × 5.5 μm |
| Image Size | 1024 × 1024 pixels (12.288 mm × 12.288 mm) | 2048 × 2048 pixels (11.264 mm × 11.264 mm) |

The calibration residuals of both prototypes were processed to compare the systematic centroid error between the output FOFP of the intensifier and the imaging chip. As shown in Figure 13, the star tracker in calibration was installed on a two-axis rotary table that could generate arbitrary rotation angles and simulate a single star entering from different directions. To rule out the other two FOFP-induced systematic errors, both sensors adopted the same type of intensifiers and were calibrated with a same commercial optical lens. To eliminate the temporal random error caused by the gain variation of the MCP, centroid data were sampled 20 times in each point. Given the difference in spatial frequency, the low frequency parts of the residuals from the three-order polynomial fitting were abandoned to exclude the spatial random error caused by fiber arrangement defection.

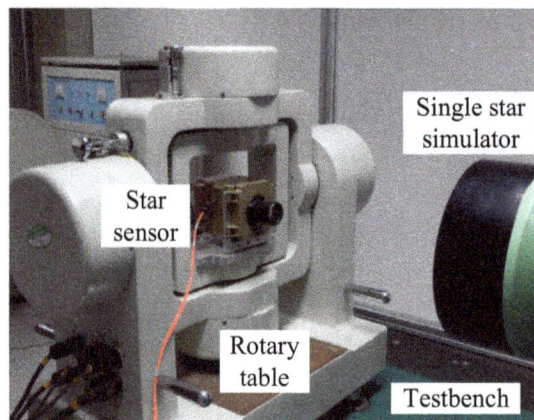

**Figure 13.** Laboratory calibration of the intensified star tracker.

The histogram of systematic centroid error magnitudes of the calibration points are shown in Figure 14. The error magnitudes of the calibration points from CMV4000 are significantly smaller than that from FTT1010M. Nearly all the error magnitudes of the points from CMV4000 is within 2 μm, whereas some of the calibration points from FTT1010M exceed 5 μm. According to Equation (45), given that the output FOFP of the fiber core diameter of the intensifier is 5.5 μm, pixel size should not exceed 4.5 μm to remain free from system centroid error. The pixel size of CMV4000 is obviously significantly closer to the set value than the pixel size of FTT1010M. The experiment result indicated that the systematic centroid error between the output FOFP of the intensifier and the imaging chip would be suppressed if the pixel size of the imaging chip matches with the fiber core diameter of the FOFP.

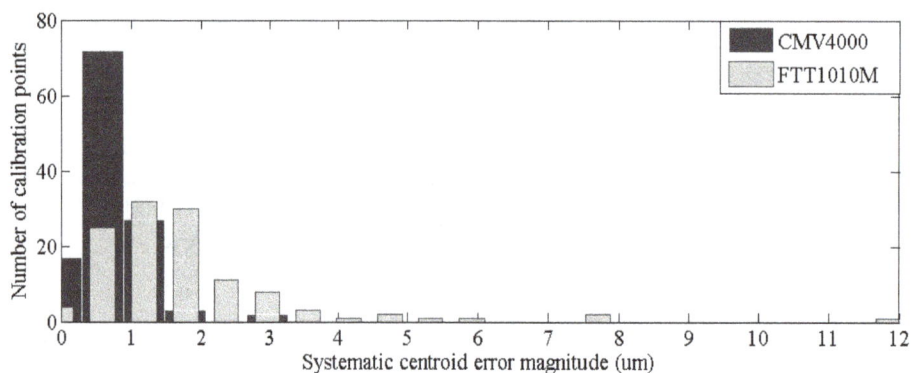

**Figure 14.** Histogram of systematic centroid errors.

## 7. Conclusions

Based on the description of the imaging process of the optoelectronic detecting system of intensified star trackers, the error types caused by an FOFP were analyzed and classified. Three possible FOFP-induced systematic centroid errors can occur through the optical path, namely: (1) the error between the optical lens and the input FOFP; (2) the error among multiple FOFPs and (3) the error between the output FOFP and the imaging chip.

The expression of centroid errors in the frequency domain was presented. According to the expression, the centroid error only depends on the optical component that the star spot passes through. If the component is isoplanatic and invariant, then the centroid error is a constant bias; otherwise, the centroid error is different from place to place. Ground on the three-step image transmission process (integration, sampling and reimaging), the PSF and MTF of the FOFP were obtained.

To reduce FOFP-induced systematic centroid error, three potential error sources were analyzed. According to the analysis, two constraints and one suggestion were obtained: (1) The circle of confusion ($6\sigma$) should be 2.49 times or higher that of the fiber interval of the input FOFP to free the star tracker from the systematic errors caused by the input FOFP; (2) The pixel size of the imaging chip should not exceed 1.64 times that of the fiber core radius of the output FOFP; (3) Adding OLPF between neighboring FOFPs was recommended to reduce the systematic error caused by multiple FOFPs. Moreover, the exact expression of the systematic centroid error caused by the FOFP was deduced.

Finally, simulations and an experiment were conducted to verify the conclusions. The error distribution of a star spot that is entering the FOFP from different positions was simulated, and the result coincides with the systematic centroid expression of the FOFP. The relation between systematic error magnitude and the ratio of the circle of confusion of the optical lens to the fiber interval of the input FOFP was simulated, and showed that the systematic error would be less than 0.1 μm when the first constraint was satisfied. Moreover, the relation between the systematic error magnitude and the core-sheath ratio was simulated. The result demonstrated that the bigger the value of the core-sheath ratio, the less systematic it would be. The processed calibration residuals regarded as the systematic error between the output FOFP and the imaging chip of two intensified star tracker prototypes were compared. The prototype that adopted a 5.5 μm-pixel-size CMOS chip had considerably less systematic errors than the prototype that adopted a 12 μm-pixel-size CCD chip when the fiber core diameter of the output FOFP was 5.5 μm. The experiment result demonstrated the correctness of the second constraint.

## Acknowledgments

This research is supported by the National Natural Science Fund of China under Grant (No. 61222304) and grants from the Specialized Research Fund for the Doctoral Program of Higher Education of China (No. 20121102110032). The authors are grateful for all of the valuable suggestions received during the course of this work.

## Author Contributions

Kun Xiong focused mainly on the theoretical derivation and paperwork. Jie Jiang was in charge of the simulation and experiment part of the study.

## Conflicts of Interest

The authors declare no conflict of interest.

## References

1. Eisenman, A.R.; Liebe, C.C. The advancing state-of-the-art in second generation star trackers. In Proceedings of the 1998 IEEE Aerospace Conference, Snowmass at Aspen, CO, USA, 21–28 March 1998; pp. 111–118.
2. Steyn, W.; Jacobs, M.; Oosthuizen, P. A high performance star sensor system for full attitude determination on a microsatellite. In Proceedings of the Workshop on Control of Small Spacecraft at the 1997 Annual AAS Guidance and Control Conference, Breckenridge, CO, USA, 5 February 1997.
3. Liebe, C.C.; Alkalai, L.; Domingo, G.; Hancock, B.; Hunter, D.; Mellstrom, J.; Ruiz, I.; Sepulveda, C.; Pain, B. Micro APS based star tracker. In Proceedings of the 2002 IEEE Aerospace Conference Proceedings, Pasadena, CA, USA, 9–16 March, 2002; Volume 2285.
4. Liebe, C.C.; Gromov, K.; Meller, D.M. Toward a stellar gyroscope for spacecraft attitude determination. *J. Guid. Control Dyn.* **2004**, *27*, 91–99.
5. Katake, A.; Bruccoleri, C. StarCam SG100: A high-update rate, high-sensitivity stellar gyroscope for spacecraft. *IS&T/SPIE Electron. Imaging* **2010**, doi:10.1117/12.839107.
6. Zhang, H.; Li, J. The effects of APS star tracker detection sensitivity. In Proceedings of the International Conference of Optical Instrument and Technology, Beijing, China, 16–19 November 2008; pp. 71601F–71608F.
7. Shen, J.; Zhang, G.; Wei, X. Simulation analysis of dynamic working performance for star trackers. *JOSA A* **2010**, *27*, 2638–2647.
8. Wei, X.; Tan, W.; Li, J.; Zhang, G. Exposure Time Optimization for Highly Dynamic Star Trackers. *Sensors* **2014**, *14*, 4914–4931.
9. Player, M. Spread functions and modulation transfer functions of fiber-optic bundles. *J. Mod. Opt.* **1988**, *35*, 1363–1372.
10. Barnard, K.J.; Boreman, G.D. Modulation transfer function of hexagonal staring focal plane arrays. *Opt. Eng.* **1991**, *30*, 1915–1919.
11. Zhang, W.; Wang, Y.; Tian, W.; Ma, W.; Zhang, H.; Sun, A. Novel methods for measuring modulation transfer function for fiber optic taper. In Proceedings of the ICO20: Optical Design and Fabrication, Changchun, China, 31 January, 2006; pp. 603413–603417.
12. Li, H.; Chen, B.; Feng, K.; Ma, H. Modulation transfer function measurement method for fiber optic imaging bundles. *Opt. Laser Technol.* **2008**, *40*, 415–419.
13. Yang, X.; Tang, Y.; Liu, K.; Liu, H.; Gao, H.; Li, Q. Modulation transfer function of partial gating detector by liquid crystal auto-controlling light intensity. In Proceedings of the Photonics and Optoelectronics Meetings, Wuhan, China, 11 December 2008; pp. 72790Z–72798Z.

14. Liebe, C.C. Accuracy performance of star trackers-a tutorial. *IEEE Trans. Aerosp. Electron. Syst.* **2002**, *38*, 587–599.

15. Yang, J.; Liang, B.; Zhang, T.; Song, J. A novel systematic error compensation algorithm based on least squares support vector regression for star sensor image centroid estimation. *Sensors* **2011**, *11*, 7341–7363.

16. Katake, A.B. Modeling, Image Processing and Attitude Estimation of High Speed Star Sensors. Ph.D. Thesis, Texas A&M University, College Station, TX, USA, 2006.

17. Alexander, B.F.; Ng, K.C. Elimination of systematic error in subpixel accuracy centroid estimation. *Opt. Eng.* **1991**, *30*, 1320–1331.

# 7

# Investigation of Two Novel Approaches for Detection of Sulfate Ion and Methane Dissolved in Sediment Pore Water Using Raman Spectroscopy

**Zengfeng Du [1], Jing Chen [1], Wangquan Ye [1], Jinjia Guo [1], Xin Zhang [2] and Ronger Zheng [1,***

[1] Optics and Optoelectronics Laboratory, Ocean University of China, Qingdao 266100, China;
E-Mails: mofeng212@163.com (Z.D.); chenjingcj40@163.com (J.C.); jxyewaqu@163.com (W.Y.);
opticsc@ouc.edu.cn (J.G.)

[2] Key Lab of Marine Geology and Environment, Institute of Oceanology, Chinese Academy of Sciences,
Qingdao 266071, China; E-Mail: xzhang@qdio.ac.cn

* Author to whom correspondence should be addressed; E-Mail: rzheng@ouc.edu.cn

Academic Editor: Mark A. Arnold

---

**Abstract:** The levels of dissolved sulfate and methane are crucial indicators in the geochemical analysis of pore water. Compositional analysis of pore water samples obtained from sea trials was conducted using Raman spectroscopy. It was found that the concentration of $SO_4^{2-}$ in pore water samples decreases as the depth increases, while the expected Raman signal of methane has not been observed. A possible reason for this is that the methane escaped after sampling and the remaining concentration of methane is too low to be detected. To find more effective ways to analyze the composition of pore water, two novel approaches are proposed. One is based on Liquid Core Optical Fiber (LCOF) for detection of $SO_4^{2-}$. The other one is an enrichment process for the detection of $CH_4$. With the aid of LCOF, the Raman signal of $SO_4^{2-}$ is found to be enhanced over 10 times compared to that obtained by a conventional Raman setup. The enrichment process is also found to be effective in the investigation to the prepared sample of methane dissolved in water. By $CCl_4$ extraction, methane at a concentration below 1.14 mmol/L has been detected by conventional Raman spectroscopy. All the obtained results suggest that the approach proposed in this paper has great potential to be developed as a sensor for $SO_4^{2-}$ and $CH_4$ detection in pore water.

**Keywords:** Raman spectroscopy; sulfate ion; methane; LCOF; CCl₄ extraction

---

## 1. Introduction

Seafloor sediments constitute some of the most extreme environments ever known [1,2], and are characterized by low temperatures, high pressures and little oxygen. Due to the oxic conditions that prevail in the world's oceans, the dominant sulfur species in seawater is the sulfate ion ($SO_4^{2-}$), which with a concentration of 29 mmol/L (2.71 g/kg) in seawater, the second most abundant anion [3], and it moves from the oceans to the sediments via various mechanisms. This makes marine sediments the main sink for seawater sulfate. Sulfate in marine sediments participates in the degradation of organic matter as a dominant electron acceptor until it is exhausted in the deeper subsurface sediment [4] where methanogenesis becomes the main terminal pathway of organic carbon mineralization [5]. Methane, as a stable end product of organic carbon mineralization, is produced exclusively by anaerobic archaea [6], and accumulates in subsurface sediments. Strong gradients in dissolved sulfate ion with depth are frequently observed, especially in the pore water that bathes gas hydrates. Methane slowly diffuses up to the sulfate zone, which is referred to as "sulfate-methane transition" (SMT), and reacts with sulfate in pore water. The coupled sulfate-methane reaction equation is $CH_4 + SO_4^{2-} \rightarrow HCO_3^- + H_2S + H_2O$, and both the methane and sulfate are consumed to depletion. The sulfate concentration gradient in pore water can be taken as a universal indicator of depth to the sulfate-methane interface (SMI) [7], which is a fundamental biogeochemical redox boundary in methane-rich and methane-gas-hydrate-bearing marine sediments [8–11]. The anomaly of methane concentration is also regarded as evidence of the existence of natural gas hydrates [9].

Much of the geochemical knowledge sought does not come from the sediments themselves but from the pore waters that contain the signature of the reactions at work [12]. Geochemical studies in the deep ocean have traditionally relied upon sample recovery by bottles, cores, and dredges deployed from surface ships, or collected by manned submersibles and remotely operated vehicles (ROVs), to provide specimens for ship or shore based analysis [13]. Each year hundreds of ocean sediment cores are taken on purpose for geochemistry analysis [14]. However, it has been found that the methane concentrations differ greatly (up to $10^3$) in conventionally recovered cores and pore water sampling with pressurized core recovery [15]. Thus an *in situ* technique for geochemistry analysis is highly desirable.

Commercial methane sensors have been developed and used for methane monitoring and underwater detection. METS, as an electrochemical sensor, has been widely used for the detection of methane [16,17]. HydroC (Contros GmbH, Kiel, Germany) is also a commercial methane sensor, which is based on direct IR absorption spectroscopy [18]. Due to the stipulation that the targets must be gaseous methane, a gas-permeable membrane is indispensable for the two methane sensors, which limits the underwater application of the sensors.

Raman spectroscopy is regarded as a powerful technique for the geochemical analysis of pore water. This is especially true in the study of the oceanic gas hydrates near the seafloor. However, although the challenges of carrying out *in situ* Raman spectroscopy detection are formidable [13], in recent years, Raman spectroscopy applications for *in situ* detection have become increasingly popular.

While in most instances, the concentration of methane in sea water is too low to be detected for conventional Raman spectroscopy, technologies for improving the sensitivity of Raman spectroscopy have been developed to broaden its underwater applications. In this paper, investigations for the improvement of the limit of detection for sulfate ion and methane dissolved in pore water using Raman spectroscopy have been carried out. Samples have been prepared in the laboratory (sodium sulfate solutions and saturated aqueous solution of methane) and pore water samples have been squeezed from sediment cores as samples for analysis. The potential application of Raman spectroscopy technology, based on the approaches proposed in this paper for *in situ* detection of sulfate ion and methane dissolved in pore water, is also discussed in this paper.

## 2. Methods

### 2.1. The Principle of Raman Signal Enhancement

The intensity of a solute's Raman signal in water can be described by Equation (1):

$$R = KI\sigma PC \tag{1}$$

where $R$ is the intensity of Raman signal, $K$ is a coefficient that is determined by the spectra acquisition system, $I$ is the excitation laser power, $\sigma$ is the Raman cross-section of the samples under investigation, $P$ is the effective optical path length, and $C$ is the molecular density of the sample [19]. $K$, $I$ and $\sigma$ are determined by the experimental setup, while $P$ can be improved in order to enhance the Raman signals of the samples. Due to the total internal reflection, the excitation laser is confined in the LCOF because of the total internal reflection, and the effective optical path length ($P$) is significantly enhanced [20]. Thus, a better sensitivity can be achieved for Raman spectroscopy.

### 2.2. Instrumental Setup

A specific Raman spectroscopy setup with LCOF is established using commercially available components. The schematic diagram of experimental setup is presented in Figure 1. A diode-pumped, solid state laser that emits at 532 nm and outputs power at 300 mW is used as the light source (LMX-532S, from Oxxius, Lannion, France). The dichroic mirror in the dotted box is detachable in order to obtain Raman spectra of the samples using the conventional Raman spectroscopy experimental setup as well as the Raman experimental setup based on LCOF.

### 2.3. Sampling

The pore water samples were acquired in sea trials of "Science III", a research vessel that belongs to the Chinese Academy of Sciences Institute of Oceanology. A 2 m length sediment core was taken from the seafloor at a depth of 53 m in North Yellow Sea basin (E 122°40′, N 38°46′), and cut into four pieces. The pore water samples were squeezed from different sediment pieces using an improved pore water sampler [21], and taken back to laboratory for Raman spectroscopy analysis.

In order to conduct quantitative analysis of $SO_4^{2-}$, a series of $Na_2SO_4$ solutions are prepared in 250 mL volumetric flasks, with 2, 5, 10, 15, 20, 25, 30, 40, 50, and 60 mmol/L of $SO_4^{2-}$ respectively. The solutions are transferred into 5 mL cuvette for acquisition of Raman spectra in the conventional way when the dichroic mirror is placed in the optical path, and pumped into the LCOF for acquisition of Raman spectra when the dichroic mirror is removed from the optical path.

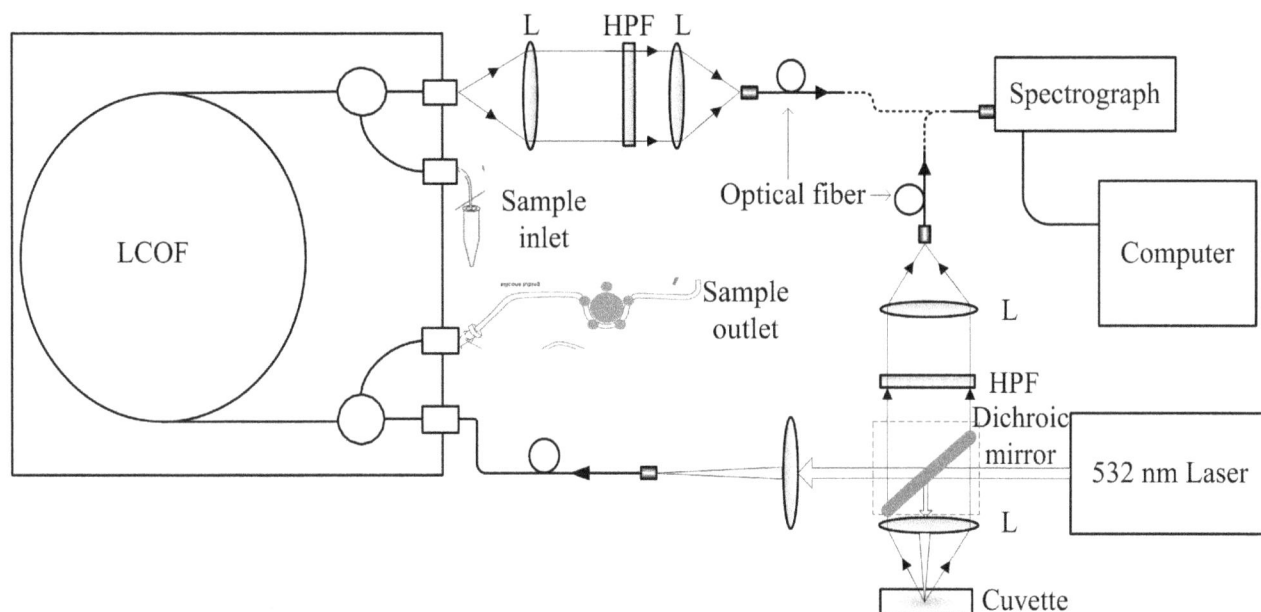

**Figure 1.** Schematic diagram of the LCOF-Raman experimental setup (R, dichroic mirror; L, optical lens; HPF, high pass filter).

To prepare a saturated aqueous solution of $CH_4$ under laboratory conditions (the concentration of methane in water is about 1.14 mmol/L), $CH_4$ is pumped into deionized water (DI water) for 1 h. During the enrichment process, $CCl_4$ is injected into the saturated aqueous solution of $CH_4$ and stirred for 0.5 h. The solution is then left for 0.5 h so the water and $CCl_4$ is separated in order to get the $CCl_4$ solution after extraction. A magnetic stirring device (IKA-RCT basic model, IKA, Aachen, Germany) is employed to make the dissolution and extraction more efficient.

*2.4. Spectra Acquisition*

For each sample, 10 spectra are recorded and averaged for analysis. The background and the dark current are measured and automatically subtracted from each subsequent spectrum. The laser power is set at 0.3 W. All the spectra are processed in Origin 8.1.

## 3. Results and Discussion

*3.1. Composition Analysis of Pore Water Samples with Different Depth*

A compact spectrometer (QE65000 Pro, from Ocean Optics, Dunedin, FL, USA) is used to acquire Raman spectra of pore water samples with an integration time of 1 s, and the original Raman spectrum of the pore water samples is shown in Figure 2. The results show that the Raman peak of $SO_4^{2-}$ at 981 $cm^{-1}$ is obviously detected, while the expected Raman peak of methane at 2917 $cm^{-1}$ cannot be detected. A possible reason for this s that the methane dissolved in pore water has escaped after sampling and the concentration of the remaining methane is too low to be detected.

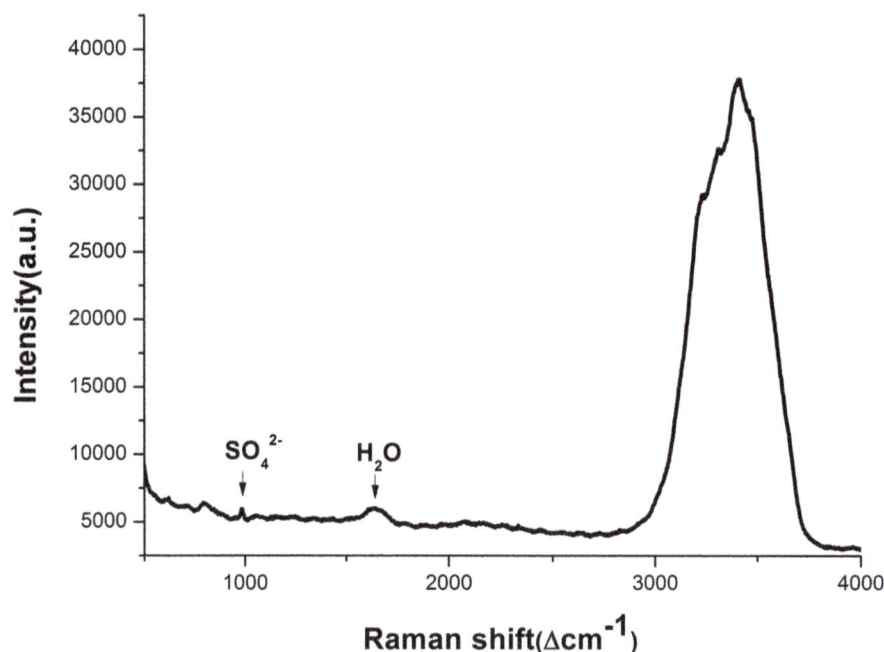

**Figure 2.** Typical Raman spectrum of the pore water samples.

The Raman spectra of the Na₂SO₄ solutions prepared in laboratory are acquired, and the internal standard normalization method is used in data processing. The Raman peak of $SO_4^{2-}$ (located at 981 cm$^{-1}$) is normalized with the Raman peak of water molecular (located at 1640 cm$^{-1}$) in this paper. The calibration curve is shown in Figure 3, and the fitted linear function is $R^* = 0.012C + 0.534$, where $R^*$ is the normalized Raman intensity of $SO_4^{2-}$, $C$ is the concentration of the $SO_4^{2-}$.

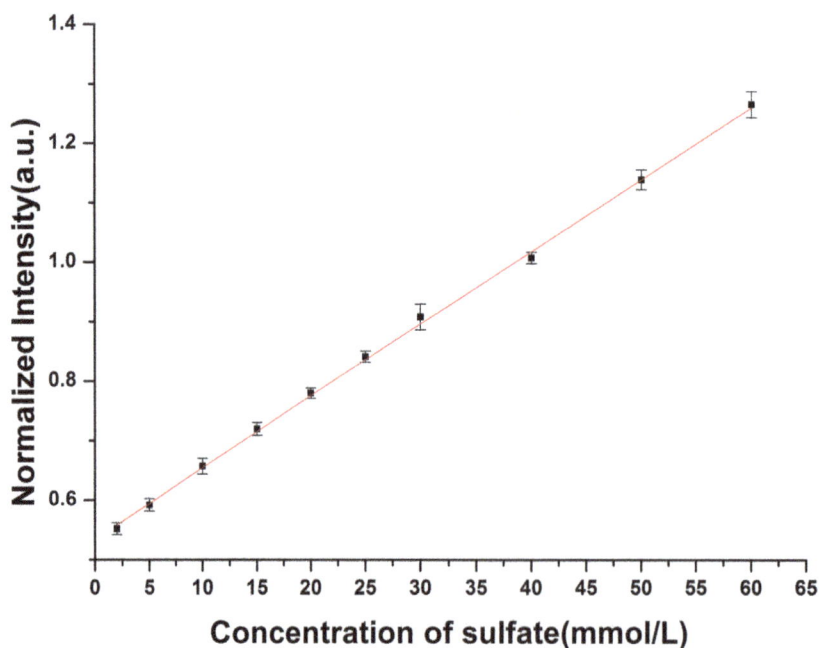

**Figure 3.** The calibration curve of sulfate concentration with conventional Raman spectroscopy.

Then the fitted linear function derived from the calibration curve, is used for quantitative analysis of the sulfate ion dissolved in pore water. According to the fitted linear function, concentrations of $SO_4^{2-}$ in pore water samples are obtained. The concentrations of $SO_4^{2-}$ in pore water are also measured by liquid chromatography. The results are demonstrated in Table 1, and the profiles of the $SO_4^{2-}$ concentrations in pore water is presented in Figure 4.

**Table 1.** Concentrations of $SO_4^{2-}$ in pore water samples measured by Raman spectroscopy and liquid chromatography.

| Samples | Sampling Depth (cm) | Concentrations of $SO_4^{2-}$ (mmol/L) | | Relative Deviation $\mid C_{Raman} - C_{LC} \mid / C_{LC}$ |
| --- | --- | --- | --- | --- |
| | | $C_{Raman}$ | $C_{LC}$ | |
| A-12-a | 20–60 | 27.1 | 28.5 | 4.91% |
| B-12-a | 60–100 | 26.1 | 23.4 | 11.54% |
| C-12-a | 100–140 | 25.1 | 23.6 | 6.36% |
| D-12-a | 140–180 | 23.1 | 22.7 | 1.76% |

$C_{Raman}$ concentrations measured by Raman spectroscopy; $C_{LC}$ concentrations measured by liquid chromatography.

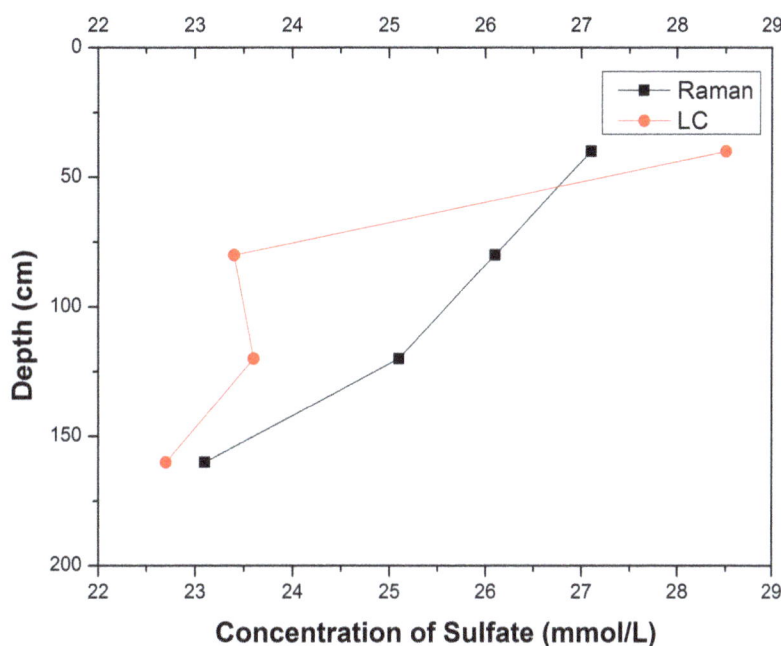

**Figure 4.** Profiles of $SO_4^{2-}$ concentrations in pore water measured by Raman spectroscopy and LC.

It can be seen that there is no significant difference between the concentrations of $SO_4^{2-}$ measured by Raman spectroscopy and LC, and the minimum relative deviation can reach 1.76%. Figure 5 also shows that the concentration of $SO_4^{2-}$ in pore water decreases as the depth increases, which indicates the existence of sulfate reduction in sediments. Because of the sulfate-methane reaction, the sulfate could be exhausted in the deeper sediment [5]. Furthermore, the concentration of $SO_4^{2-}$ would be too low to be detected for Raman spectroscopy. Thus, an enhancement technology for Raman signal of $SO_4^{2-}$ is highly desired for geochemical analysis of pore water.

## 3.2. Enhancement for Raman Signal of $SO_4^{2-}$ with LCOF

It is reported that up to a 100-fold improvement of Raman signals can be observed using the LCOF-Raman experimental setup compared to the conventional Raman experimental setup [22], and the performance of different LCOF geometries differs greatly [23].

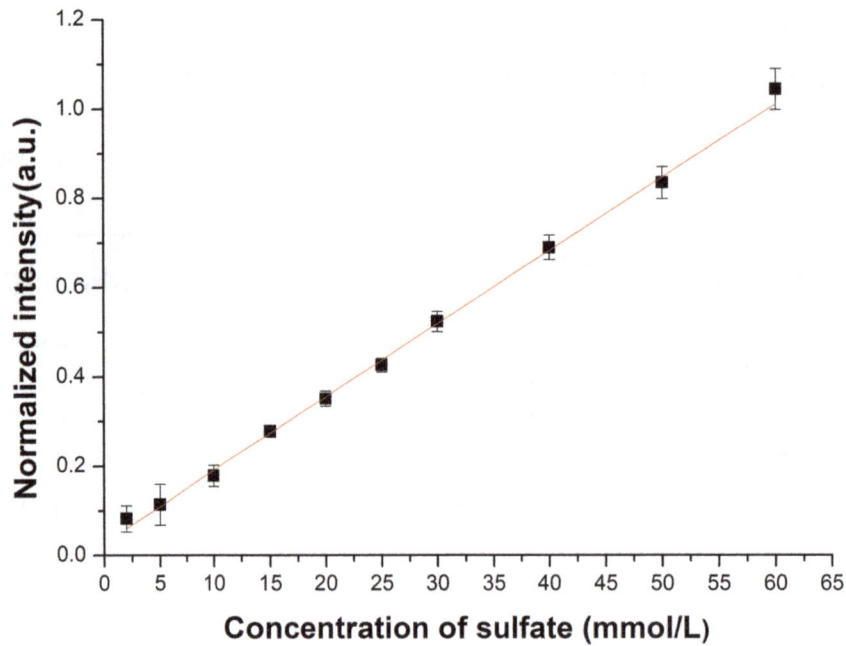

**Figure 5.** The calibration curve of sulfate concentration with LCOF-Raman spectroscopy.

A special LCOF-Raman experimental setup is established in laboratory. The physical length of the LCOF is about 100 cm (LWCC-2100, from Ocean Optics), and the spectrometer is a QE65000 Pro (from Ocean Optics) with an integration time of 1 s for each measurement. Raman spectra of the $Na_2SO_4$ solutions prepared in laboratory are also acquired by the LCOF-Raman experimental setup. The internal standard normalization method is used in data processing. The calibration curve is shown in Figure 5, and the fitted linear function is $R^* = 0.016C + 0.027$. Table 2 presents the limit of detection (LOD) of the conventional Raman and the LCOF-Raman experimental setups.

**Table 2.** The fitted linear functions and LODs of conventional Raman and LCOF-Raman experimental setup. (QE65000 is employed as spectrometer, and the length of the LCOF is 100 cm.)

|  | Fitted Linear Function | LOD (mmol/L) |
|---|---|---|
| Conventional Raman setup | $R^* = 0.012C + 0.534$ | 2.15 |
| LCOF-Raman setup | $R^* = 0.016C + 0.027$ | 1.50 |

Figure 6 shows the Raman spectra of 30 mmol/L solution of $Na_2SO_4$ using both the LWCC-2100 and a cuvette as sample container. The solid line is the Raman spectrum acquired by the LCOF-Raman (LWCC-2100) setup, while the dashed-dot line is the Raman spectrum acquired by the conventional Raman setup as a comparison. The backgrounds are subtracted for both spectra. The comparison of the spectra shows an increase in the Raman signal using the LCOF: the peak intensity and area are amplified by approximately two fold.

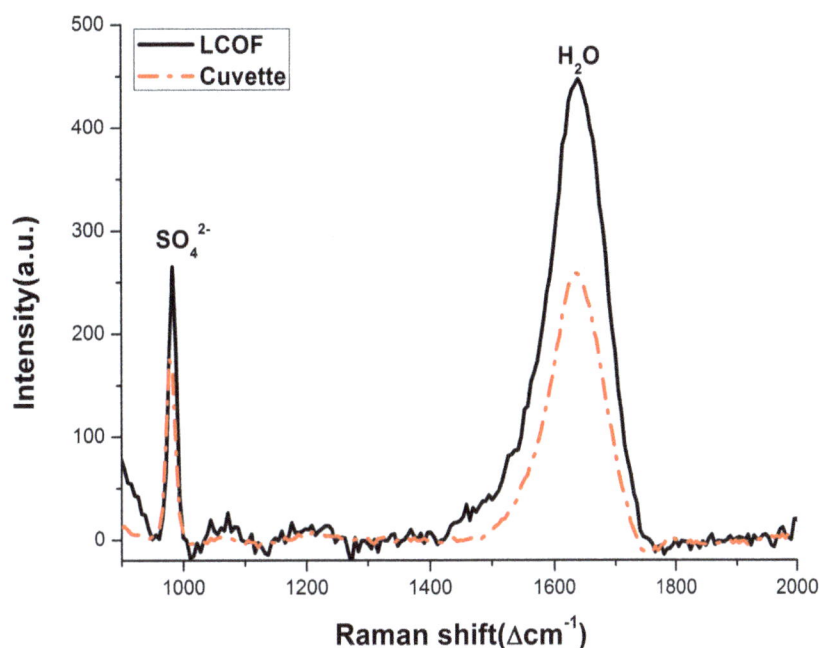

**Figure 6.** Raman spectra of 30 mmol/L sulfate solution using the LCOF-Raman experimental (LWCC-2100) setup and conventional experimental setup, respectively. The Raman peak of $SO_4^{2-}$ is located at 981 $cm^{-1}$, and the Raman peak of $H_2O$ is located at 1640 $cm^{-1}$.

In another updated setup, a LCOF (LWCC-3050, from Ocean Optics) with the same samples ($Na_2SO_4$ water solutions prepared in laboratory) pumped in and with 50 cm physical length, has been chosen for comparison. To achieve a better performance, a more sensitive spectrometer (PPO Raman, from P&P Optica, Waterloo, ON, Canada) together with a 2000 × 256 back illuminated CCD (DU416A-LDC-DD, from Andor Technology, Belfast, UK) is used. The spectra are averaged from ten measurements taken with an integration time of 0.1 s for each measurement. Then the calibration curves of sulfate concentration with the conventional Raman and LCOF-Raman spectroscopy can be obtained.

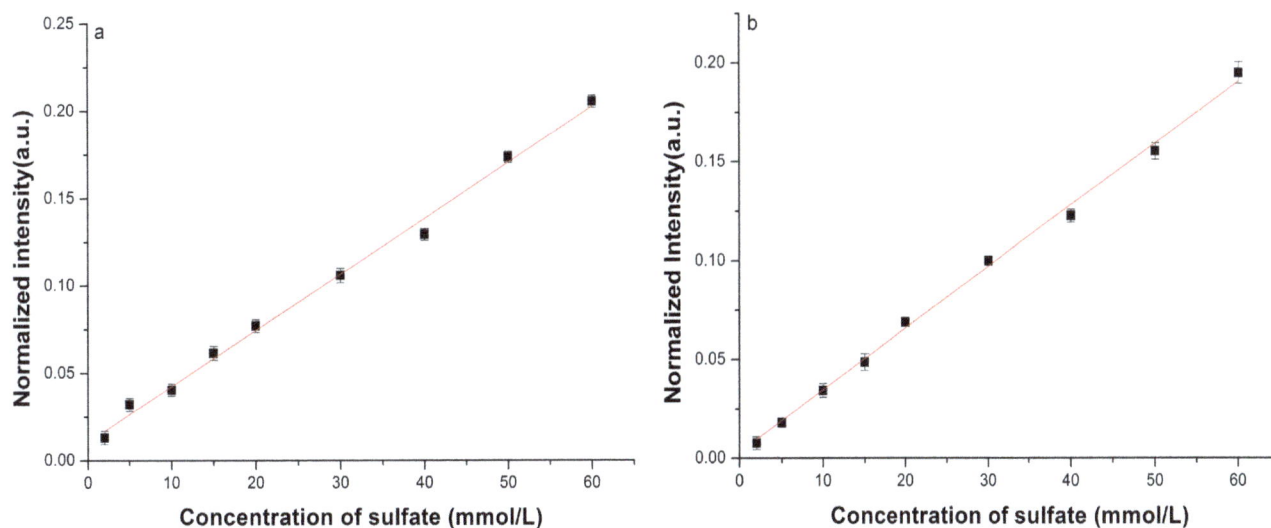

**Figure 7.** The calibration curve of sulfate concentration with (**a**) conventional Raman spectroscopy and (**b**) LCOF-Raman spectroscopy.

**Table 3.** The fitted linear functions and LODs of conventional Raman and LCOF-Raman experimental setup. (PPO Raman is employed as spectrometer, and the length of the LCOF is 50 cm.)

|  | Fitted Linear Function | LOD (mmol/L) |
|---|---|---|
| Conventional Raman setup | $R^* = 0.0032C + 0.010$ | 1.80 |
| LCOF-Raman setup | $R^* = 0.0031C + 0.0034$ | 0.35 |

Figure 7 shows the calibration curves, and Table 3 presents the LOD of the conventional Raman and the LCOF-Raman experimental setups. Figure 8 shows the typical Raman spectra obtained using both LWCC-3050 and cuvette as sample containers, with solid line and dash-dot line represented respectively. It can be seen from Figure 8 that the Raman signal of $SO_4^{2-}$ obtained with the LCOF-Raman setup is much higher than that with conventional Raman setup. Over 10-fold enhancement is achieved with the LCOF approach.

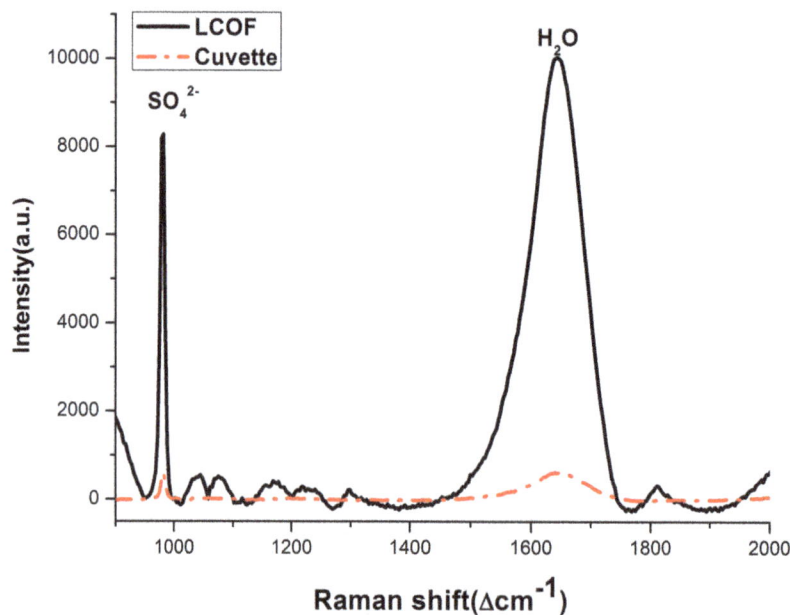

**Figure 8.** Raman spectra of 30 mmol/L sulfate solution using the LCOF-Raman experimental (LWCC-3050) setup and conventional experimental setup respectively.

Tables 2 and 3 present the LODs of the Raman experimental setups, and it is found that the LOD of Raman spectroscopy can be improved with LCOF assistance. The obtained results indicate that the physical length of the LCOF has a great influence on the enhancement of the Raman signals. It seems that the 50 cm LCOF has a better Raman signal enhancement than the 100 cm LCOF.

### 3.3. Dissolved Methane Detection Aided with an Enrichment Process

As is known, the maximum concentration of dissolved methane is about 1.14 mmol/L (the concentration of its saturated solution under laboratory conditions), which is still too low to be detected by the LCOF-Raman (LWCC-3050) experimental setup. In order to detect the methane dissolved in water, an approach based on $CCl_4$ extraction is introduced in this work. $CCl_4$ is chosen as an extraction agent for two reasons: the solubility of $CH_4$ in $CCl_4$ is much larger than that in $H_2O$ and

CCl₄ is immiscible with water. Thus, the trace $CH_4$ dissolved in water is enriched into $CCl_4$ after the extraction. Then, the $CCl_4$ after extraction is taken as samples for Raman spectrum acquisition. The molecular density of $CH_4$ ($C$ in Equation (1)) in $CCl_4$ is much bigger than that in $H_2O$. Finally, the Raman spectrum of $CCl_4$ after the extraction acquired by the conventional experimental setup with an integration time of 0.1 s, which is an average result of ten measurements, is shown in Figure 9.

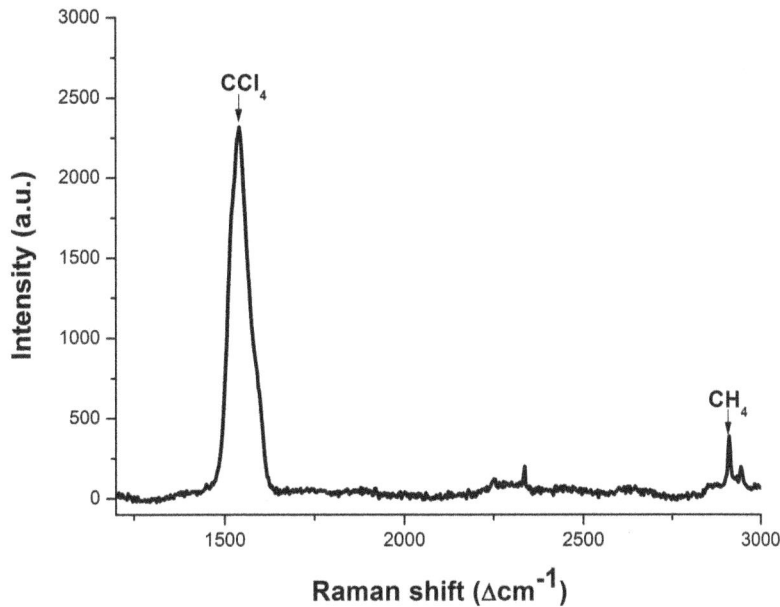

**Figure 9.** Raman spectrum of CCl₄ after extraction after conventional experimental setup.

The Raman peak of methane is located at 2907 cm$^{-1}$ and can be clearly observed in Figure 9. This indicates that, after the extraction, the methane dissolved in water is enriched into the $CCl_4$. The preliminary result shows that the methane dissolved in water with concentration below 1.14 mmol/L could be indirectly detected assisted by $CCl_4$ extraction. There are still many opportunities for optimizing this approach, and there is a long way to go in order to achieve a quantitative analysis of methane dissolved in water.

## 4. Conclusions

For the compositional analysis of pore water, Raman spectra of the pore water samples obtained from the sea trials were acquired. According to the linear function obtained from the calibration curve, the concentration of $SO_4^{2-}$ in pore water is inversely calculated according to the linear function obtained from the calibration curve. It is found that the concentration of $SO_4^{2-}$ in pore water samples decreases as the depth increases. However, methane cannot be detected using Raman spectroscopy because of its low concentration. Two approaches are proposed and used for a better analysis of pore water samples. One approach uses a LCOF as a sample container to enlarge the optical path length for detection of $SO_4^{2-}$. The other approach is an enrichment process for methane with $CCl_4$ extraction. With the assistance of a LCOF whose physical length is 50 cm, the LOD of Raman spectroscopy is significantly improved, and the Raman signal of $SO_4^{2-}$ is amplified over 10 times compared to that obtained with a conventional Raman experimental setup. By the means of extraction, the trace methane dissolved in water is enriched into $CCl_4$ because the solubility of methane in water and $CCl_4$ differs immensely. Hence the methane dissolved in water with concentration below 1.14 mmol/L could be indirectly detected. There are still opportunities to optimize the performance of these two approaches. Furthermore, all of the obtained results suggest that the proposed approaches in this paper have great potential to be developed to a sensor for detection of sulfate ion and methane dissolved in sediment pore water.

## Acknowledgments

The work was supported by National High Technology Research and Development Program of China (Grant No. 2012AA09A405) and Natural Science Foundation of China (Grant No. 41006059).

## Author Contributions

Ronger Zheng supervised the research project and revised the draft. Zengfeng Du and Jinjia Guo proposed the idea. Xing Zhang provided the pore water samples and discussed the results. Zengfeng Du and Wangquan Ye performed the experiments. Jing Chen processed the data. Zengfeng Du wrote the manuscript.

## Conflicts of Interest

The authors declare no conflict of interest.

## References

1.  Prieur, D.; Erauso, G.; Jeanthon, C. Hyperthermophilic life at deep-sea hydrothermal vents. *Planet. Space Sci.* **1995**, *43*, 115–122.

2.  Simoneit, B.R.T.; Schoell, M.; Dias R.F.; de Auuino Neto, F.R. Unusual carbon isotope compositions of biomarker hydrocarbons in a Permian tasmanite. *Geochim. Cosmochim. Acta* **1993**, *57*, 4205–4211.

3.  Vairavamurthy, M.A.; Orr, W.L.; Manowitz, B. Geochemical transformations of sedimentary sulfur: An introduction. In *ACS Symposium Series*; American Chemical Society: Washington, DC, USA, 1995; pp. 1–15.

4.  Henrichs, S.M.; Reeburgh, W.S. Anaerobic mineralization of marine sediment organic matter: Rates and the role of anaerobic processes in the oceanic carbon economy. *Geomicrobiol. J.* **1987**, *5*, 191–237.

5.  Jorgensen, B.B.; Kasten, S. Sulfur Cycling and Methane Oxidation. In *Marine Geochemistry*; Horst, D.S., Matthias, Z., Eds.; Springer: Berlin/Heidelberg, Germany, 2006; pp. 271–309.

6.  Whitman, W.B.; Bowen, T.L.; Boone, D.R. The methanogenic bacteria. In *The Prokaryotes*; Dworkin, M., Balows, A., Eds.; Springer US: New York, NY, USA, 2006; pp. 165–207.

7.  Borowski, W.S.; Paull, C.K.; Ussler, W. Carbon cycling within the upper methanogenic zone of continental rise sediments; an example from the methane-rich sediments overlying the Blake Ridge gas hydrate deposits. *Mar. Chem.* **1997**, *57*, 299–311.

8.  Claypool, G.E.; Kaplan, I.R. The origin and distribution of methane in marine sediments. In *Natural Gases in Marine Sediments*; Isaac, R.K., Ed.; Springer US: New York, NY, USA, 1974; pp. 99–139.

9. Borowski, W.S.; Paull, C.K.; Ussler, W. Marine pore-water sulfate profiles indicate *in situ* methane flux from underlying gas hydrate. *Geology* **1996**, *24*, 655–658.

10. Borowski, W.S.; Paull, C.K.; Ussler, W. Global and local variations of interstitial sulfate gradients in deep-water, continental margin sediments: Sensitivity to underlying methane and gas hydrates. *Mar. Geol.* **1999**, *159*, 131–154.

11. Ussler, W.; Paull, C.K. Rates of anaerobic oxidation of methane and authigenic carbonate mineralization in methane-rich deep-sea sediments inferred from models and geochemical profiles. *Earth Planet. Sci. Lett.* **2008**, *266*, 271–287.

12. Zhang, X.; Walz, P.M.; Kirkwood, W.J.; Hester, K.C.; Ussler, W.; Peltzer, E.T.; Brewer, P.G. Development and deployment of a deep-sea Raman probe for measurement of pore water geochemistry. *Deep Sea Res. Part I Oceanogr. Res. Pap.* **2010**, *57*, 297–306.

13. Brewer, P.G.; Malby, G.; Pasteris, J.D.; White, S.N.; Peltzer, E.T.; Wopenka, B.; Freeman, J.; Brown, M.O. Development of a laser Raman spectrometer for deep-ocean science. *Deep Sea Res. Part I Oceanogr. Res. Pap.* **2004**, *51*, 739–753.

14. Reeburgh, W.S. Oceanic Methane Biogeochemistry. *Chem. Rev.* **2007**, *107*, 486–513.

15. Paull, C.K.; Ussler, W. History and Significance of Gas Sampling During DSDP and ODP Drilling Associated with Gas Hydrates. In *Natural Gas Hydrates: Occurrence, Distribution, and Detection*; Paull, C.K., Dillon, W.P., Eds.; American Geophysical Union: Washington, DC, USA, 2001; pp. 53–65.

16. Newman, K.R.; Cormier, M.H.; Weissel, J.K.; Driscoll, N.W.; Kastner, M.; Solomon, E.A.; Robertson, G.; Hill, J.C.; Singh, H.; Camilli, R.; *et al.* Active methane venting observed at giant pockmarks along the US mid-atlantic shelf break. *Earth Planet. Sci. Lett.* **2008**, *267*, 341–352.

17. Marinaro, G.; Etiope, G.; Bue, N.L.; Favali, P.; Papatheodorou, G.; Christodoulou, D.; Furlan, F.; Gasparoni, F.; Ferentinos, G.; Masson, M.; *et al.* Monitoring of a methane-seeping pockmark by cabled benthic observatory (Patras Gulf, Greece). *Geo-Mar. Lett.* **2006**, *26*, 297–302.

18. Fietzek, P.; Kramer, S.; Esser, D. Deployments of the HydroC™($CO_2$/$CH_4$) on stationary and mobile platforms-Merging trends in the field of platform and sensor development. In Proceedings of the IEEE OCEANS, Waikoloa, HI, USA, 19–22 September 2011; pp. 1–9.

19. Pelletier, M.J. *Analytical Applications of Raman Spectroscopy*; Blackwell Science Ltd.: Osney Mead, Oxford, UK, 1999; p. 478.

20. Altkorn, R.; Koev, I.; van Duyne, R.P.; Litorja, M. Low-loss liquid-core optical fiber for low-refractive-index liquids fabrication, characterization, and application in Raman spectroscopy. *Appl. Opt.* **1997**, *36*, 8892–8898.

21. Reeburgh, W.S. An improved interstitial water sampler. *Limnol. Oceanogr.* **1967**, *12*, 163–165.

22. Qi, D.; Berger, A.J. Quantitative analysis of Raman signal enhancement from aqueous samples in liquid core optical fibers. *Appl. Spectrosc.* **2004**, *58*, 1165–1171.

23. Altkron, R.; Malinsky, M.D.; van Duyne, R.P.; Koev, I. Intensity considerations in liquid core optical fiber Raman spectroscopy. *Appl. Spectrosc.* **2001**, *55*, 373–381.

# A Novel Software Architecture for the Provision of Context-Aware Semantic Transport Information

**Asier Moreno [1], Asier Perallos [1,\*], Diego López-de-Ipiña [1], Enrique Onieva [1], Itziar Salaberria [1] and Antonio D. Masegosa [1,2]**

[1] Deusto Institute of Technology (DeustoTech), University of Deusto, Bilbao 48007, Spain;
E-Mails: asier.moreno@deusto.es (A.M.); dipina@deusto.es (D.L.-I.);
enrique.onieva@deusto.es (E.O.); itziar.salaberria@deusto.es (I.S.);
ad.masegosa@deusto.es (A.D.M.)

[2] IKERBASQUE, Basque Foundation for Science, Bilbao 48011, Spain

\* Author to whom correspondence should be addressed; E-Mail: perallos@deusto.es

Academic Editor: Jesús Fontecha

**Abstract:** The effectiveness of Intelligent Transportation Systems depends largely on the ability to integrate information from diverse sources and the suitability of this information for the specific user. This paper describes a new approach for the management and exchange of this information, related to multimodal transportation. A novel software architecture is presented, with particular emphasis on the design of the data model and the enablement of services for information retrieval, thereby obtaining a semantic model for the representation of transport information. The publication of transport data as semantic information is established through the development of a Multimodal Transport Ontology (MTO) and the design of a distributed architecture allowing dynamic integration of transport data. The advantages afforded by the proposed system due to the use of Linked Open Data and a distributed architecture are stated, comparing it with other existing solutions. The adequacy of the information generated in regard to the specific user's context is also addressed. Finally, a working solution of a semantic trip planner using actual transport data and running on the proposed architecture is presented, as a demonstration and validation of the system.

**Keywords:** Intelligent Transportation Systems; multimodal transport information; semantic middleware; Linked Open Data; context-aware computing

---

# 1. Introduction

Progress made over the last years in the application of ICT to transportation systems is extensive, constant and diverse. With regard to the software services for transport, some of the elements that have evolved the most, providing a high added value to the user, are the multimodal trip planning solutions. In this area, solutions like *Google Maps* or *OpenTripPlanner* have made important progresses in facilitating trip management and planning to the users.

Efforts are also being made at institutional level to provide the tools that allow citizens to opt for sustainable transport solutions. Thus, according to data from the International Association of Public Transport (UITP) [1], it is expected that public transport by 2025 will double its market share compared to 2009, thereby completing the transition to a sustainable transport model.

However, there is still much room for improvement in this area. Existing tools are not sufficiently interoperable due to the lack of a universal and consistent format to represent transport information. Likewise, they do not take into account relevant factors, such as the user's context. Moreover, in most cases these planning tools are closed so the access to its information becomes very costly.

This paper proposes a novel software architecture to address the above-mentioned limitations by incorporating innovative technologies such as semantic middleware, context-awareness computing or Linked Open Data, given that they have already been successfully tested in other application areas.

The article is structured into four sections. Section 2 provides an overview of the state of the art related to the technologies and knowledge areas covered in the proposed solution. Section 3 details the design and implementation of the Multimodal Transport Ontology (MTO) used for the management of transit information. Section 4 establishes the design of a distributed software architecture for semantic information provision allowing dynamic integration of transport data, detailing its components and characteristics. In Section 5, the overall system is validated by the deployment of a trip planning solution running on actual transport data. Finally, conclusions and future work derived from the experimentation analysis are given.

# 2. State of the Art

The proposed architecture is based on several areas of knowledge. This section aims to show an overview of the state of the art within these areas. The main research field will be introduced first: Advanced Traveler Information Systems (ATIS). Then, existing solutions for multimodal transport data management will be evaluated, as well as successful alternatives for semantic data provision.

## 2.1. Advanced Traveler Information Systems

Transportation systems efficiency is essential for economic development. Intelligent Transportation Systems (ITS) can be defined as a set of applications within computer science, electronics and communications that, from a social, economic and environmental point of view, are aimed at improving mobility, security and transport productivity, optimizing the use of infrastructures and energy consumption and improving the capacity of the transport systems [2].

Within the ITS field, this work is focused on providing enriched transport information to the user through the integration of information sources and the generation of new knowledge. The most relevant research works in this area are focused on ATIS, designed to assist travelers in planning their trips and route optimization [3,4].

These systems use ICT in order to collect, process and distribute the latest traffic information, road conditions, travel time, expected delays, alternative routes and/or weather conditions to the user, giving travelers the opportunity to make informed decisions about when to travel, what mode of transport to use or which route to take.

Research done in these concepts has been instrumental in the development of software tools and commercial applications for journey planning, one of the areas with greater acceptance within the ATIS. Examples of it are *Google Transit* or *Moveuskadi* in the private field or *OpenTripPlanner* as open source software.

### 2.2. Transit Information Formats and Standards

Transportation companies have their own information about service planning related to routes and schedules made by their fleets of vehicles. But as Campbell *et al.* [5] indicate on their work, the effectiveness of transportation information systems depends largely on the ability to integrate information from diverse sources and the suitability of this information to the specific user.

Currently there are initiatives for the publication and exchange of carriers' transit information, which are allowing developers to consume this information and to integrate it in their applications in an interoperable way. The two main existent solutions (as to their widespread use and community support) for transit data modelling and publishing are GTFS, from Google, and WFS from the Open Geospatial Consortium (OGC) along with several ad-hoc solutions defined by transport agencies themselves. The details of these solutions are described below.

### 2.2.1. General Transit Feed Specification (GTFS)

The aforementioned applications use GTFS as the data format for modelling transit information. GTFS defines a common format for public transportation schedules and associated geographic information, having established itself as the *de facto* standard for the representation of transit data, thanks in large part to the support received from Google and its maps services. However, the format relies on comma-separated value (CSV) files to represent the information which is then compressed and stored. This leads to isolated and outdated data that is neither easily queryable nor extensible.

### 2.2.2. Web Feature Service (WFS)

The Open Geospatial Consortium Web Feature Service Interface Standard (WFS) provides an interface allowing requests for geographical features across the web using platform independent calls. The XML-based GML furnishes the default payload encoding for transporting the geographic features.

The WFS specification defines interfaces for describing data manipulation operations of geographic features. Data manipulation operations include the ability to get or query features based on spatial and non-spatial constraints and create/update/delete new feature instances. Implementations of the WFS standard are however scarce, mainly due to the complexity of the data model and the verbosity of the required queries.

## 2.2.3. Ad-Hoc Solutions

There are also other systems to represent and/or provide transit information, mostly defined by agencies or institutions that manage their own data. One of the most prominent is *TransXChange* used in the United Kingdom and Australia to interchange bus service planning information.

Therefore, various systems coexist in order to represent transit information. WFS specification is focused to make queries on geospatial data stored in Geographic Information Systems (GIS) and its use is complex. Ad hoc solutions are optimal in their domain but they are not interoperable. GTFS is the *de facto* standard, however, the treatment and consultation of its data is not trivial, because they are not structured and do not allow the inclusion of qualitative attributes of the route (like ecological or tourist interest) which are increasingly relevant when making schedules.

Table 1 shows a schematic comparison between the aforementioned formats and MTO, the proposed solution for the management and provision of multimodal transport information. A set of features, considered relevant for data integration and interoperability are presented.

**Table 1.** Comparison between formats for transport information provision.

|                | GTFS | WFS | Ad-Hoc Solutions | MTO |
|----------------|------|-----|------------------|-----|
| **Classification** | Open Data | Open Data | Private | Open Data |
| **Structure** | CSV | GML (XML) | Variable | Formal Ontology |
| **Extensibility** | No | No | No | Yes |
| **Linkable** | No | No | No | Yes |
| **Queryable** | Programmatic | Web Service | API | Direct (SPARQL) |
| **Data Access** | Complete | Limited | Variable | Complete |

*Classification* defines the legal restrictions applied to the provided data (e.g., open access and redistributable or privative).

*Structure* refers to the format of the data. It is important to classify these formats into non machine-readable (e.g., PDF file) and machine-readable (CSV, XML, taxonomy, *etc.*)

*Extensibility* defines the intrinsic capability of the format to be extended with other relevant data thereby enriching the original data.

*Linkable* exposes the ability of the format to be included in other datasets as a reference (e.g., using URIs) allowing even greater enrichment.

*Queryable* and *Data Access* make reference to the accessibility of the data in relation to the consultation mechanism and the completeness of the information provided, respectively.

Thus, GTFS provides complete access to the data but has to be done programmatically, with a custom scraper, as no interface is provided. WFS provides an interface but the queries are predefined so the access is limited. MTO provides complete access through standard yet customizable SPARQL queries allowing the user to get the information needed, even enriched using the mechanisms provided to link the data with other relevant datasets. Section 4 extends this description with a more detailed view of the system architecture including some example queries and SPARQL specification details.

## 2.3. New Trends in Transit Information

One of the main objectives of the proposed architecture is to improve the relevance of the transport information currently provided to the user. In the last years, the efforts on user services and transportation information have been focused on data interoperability in order to obtain this value added information. Figure 1 shows the evolution of the transport information and its transformation through different processes making it more relevant for the user.

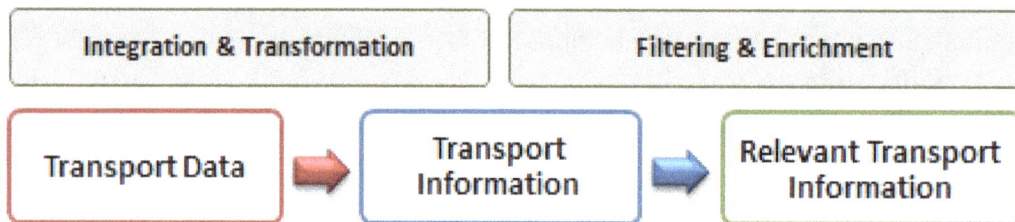

**Figure 1.** Evolution of transport information.

In order to support these processes and provide solutions to new (quantitative and qualitative) information integration, sharing and aggregation limitations, there are studies about the possibilities offered by relatively recent emerging paradigms in the computing field, such as ontologies and the Web of Data. We describe some of these studies in the next subsections.

### 2.3.1. Ontologies

The work carried out by the Artificial Intelligence community showed evidence that formal ontologies could be used to specify knowledge between different entities [6,7].

In computing, the ontology term refers to the formulation of a comprehensive and rigorous conceptual schema given within one or more domains [8]. Being independent of language and understandable by computers, ontologies are useful because they help to achieve a common and integrated understanding of descriptive information. This not only makes it easier for other human users to understand the specifically intended meaning of the models, but also means that other tools can use the definitions transparently [9].

### 2.3.2. Semantic Web

Ontologies are the heart of the Semantic Web, an extension of the World Wide Web in which the meanings (semantics) of information and services are defined [10] allowing to satisfy the requests of people and machines using web content. On the web there are millions of resource accesses, which have brought a lot of success, but also provoke problems of information overload and sources heterogeneity. The Semantic Web helps to solve these problems reducing the cognitive effort of users delegating them to agents that can reason, and make inferences to offer accurate information.

To achieve this goal of classification and provision of relevant information, Semantic Web applications used a set of standardized languages and technologies (essentially RDF, SPARQL and OWL) collaborating between them to turn the web into a global infrastructure where data and documents can be shared and reused.

## 2.3.3. Linked Data

Once established the formats (RDF, OWL) used to model specific vocabularies, thus generating ontologies for specific domains, and having defined the semantic web as an infrastructure to support these domains, the next step is to integrate this vocabularies or, in other words, link the data. The concept of Linked Open Data (LOD) arises so as the mechanism to interconnect RDF datasets available on the Internet using the HTTP protocol, as with HTML documents.

In computing, Linked Data describe a method to publish structured information, so that it can be interconnected and thus be more useful. It is based on standard web technologies such as HTTP, RDF and URIs, but instead of using them to serve web pages to people, it extends them to share information in a way that can be processed automatically by computer. As the resulting Web of Data is based on standards and a common data model, it becomes possible to implement generic applications that operate over the complete data space. This allows connecting and retrieving relevant data from various sources [11].

In order to standardize and measure the quality of the published datasets which are beginning to be huge, in 2010 a metric which is known as "*5-star Linked Data*" was introduced by Berners-Lee [12]. Basically, it consists on a set of incremental characteristics that should meet the published data to be considered as Linked Data under community criteria. The more functionality the dataset meets, the better the quality of it. Table 2 shows that classification.

**Table 2.** *5-star* Linked Data rating system.

| Stars | Description | Acronym | Example |
|---|---|---|---|
| ★ OL | Available on the web | OL: On-Line | PDF |
| ★★ OL RE | Available as machine-readable structured data | RE: Readable | XLS |
| ★★★ OL RE OF | Non-proprietary format | OF: Open Format | CSV |
| ★★★★ OL RE OF URI | Using URIs to denote things | URI: Universal Resource Identifier | RDF |
| ★★★★★ OL RE OF URI LD | Link data to related datasets | LD: Linked Data | RDF |

This concept has gained momentum after the publication of large semantic resources as DBpedia or Bio2RDF and the announcement by some governments about its decision to make public their data on a set of Open Government initiatives.

In October 2007 datasets of more than two billion triples (a data entity used for storing semantic information composed of subject-predicate-object) were counted, interlinked by means of more than two million RDF links [13]. In September 2011 this information had grown to thirty-one billion RDF triples and five hundred million links. Such efforts, both institutional and personal, are resulting in the publication of data under the characteristics and infrastructure of the Semantic Web.

## 2.4. Integrating LOD for Multimodal Transportation

Capabilities to promote knowledge sharing, structuring such knowledge and interoperability between systems have favored the use of ontologies and LOD in many areas of study and different application domains like medicine [14,15], education [16,17] or logistics [18,19]. Therefore, it should not be surprising to find similar work in the field of transportation information, motivated by various objectives, methods and expected results, but all under the premise of semantic data modeling.

### 2.4.1. Geospatial Data Management

Interoperability is becoming essential for geographic information systems whose information is usually stored in geospatial databases, accessible only via GIS. However, these sources are increasingly heterogeneous so it is necessary to consider this heterogeneity and favour methods that enable interoperability between geographic tools in order to satisfy the growing demand in the use and sharing of geospatial data [20,21].

Traditional GIS systems perform spatial queries using a method based on keywords. This approach is unable to fully express the user's needs, because of the lack of geographic concepts (semantics) in the data set. In this context, the most promising approach to end this ambiguity is the implementation of geospatial semantics using ontologies for geographic datasets.

Lorenz *et al.* made a comprehensive survey analysis in their work Ontology of Transportation Networks [22], where points out the efforts made by international institutions in order to standardize geographical information and explained the ontology of a transportation network derived from the ISO Geographic Data Files.

The thesis work presented by Lemmens [23] delves into the need to seek interoperability between different datasets and web services based on geographic information. It exposes how the paradigm change in terms of software architecture (from a centralized mode to a distributed and interconnected one) has influenced substantially in the available geographical information tools.

On the other hand, Zhao [24] presented a paper where geographical data interoperability is tackled from a distinct perspective. While the authors cited above focus their efforts on geographic information semantic modeling, designing one or more ontologies to try to get a direct translation to a richer semantic model for geospatial formats, Zhao seeks to reuse existing standard formats and geospatial protocols and enhance them using semantics in the query layer.

He argues that the direct conversion of all present geospatial data (stored mainly in geographical databases) to an ontological model is not a viable alternative because the process would be prolonged in time, being subject to the occurrence of errors and inefficient. Furthermore, existing tools for ontology management (such as Protégé) would not be able to bear such a heavy burden of instances mainly due to the required memory consumption.

## 2.4.2. Transport Information Modelling

Although several authors have tried to manage GIS systems' geographical information by ontologies, it has been proven that the complexity of this type of information, as well as its capability requirements make it impractical to compute large-scale application of such solutions. Therefore, the current approach is to direct the research towards domains or more specific areas within the transport information.

Niaraki [25] developed an ontology-based personalized route planning system using multi-criteria decision making. Another related work by Houda [26] focused on the information required by the passenger for preparing a journey, choosing the best way to move from one point to another using multimodal transportation. A work by Gunay [27] presents a more general solution, generating a semantic geoportal based on the Infrastructure for Spatial Information in the European Community (INSPIRE) data theme. The aim of this work is to investigate the use of semantics to empower the traditional GIS approach.

Along with the use of ontologies, the inclusion of contextual information also derives in the enrichment of the resulting information [28]. Examples of these are the recent solutions presented by Kim [29] or Bujan [30] for context management in the fields of healthcare and tourism, respectively.

All the research discussed above concerns ontology studies. However, each work is motivated by different objectives, methodologies or expected results. Our goal is to construct a distributed software architecture that allows, through the formalization of transit data acquired from heterogeneous sources together with the integration of relevant information, the creation of software services related to multimodal mobility.

## 3. MTO: Multimodal Transport Ontology

Section 2 highlighted the limitations encountered with the definition of a language that allow, on the one hand, to formally model transport information and, on the other, to facilitate its consultation and provision to the user. Since these features are very satisfactorily resolved through the use of ontologies and Linked Open Data as demonstrated in other areas, the establishment of an ontology as a format for the management and provision of multimodal transport information is conducted.

### 3.1. Design Methodology

A body of formally represented knowledge is based on a conceptualization: the objects, concepts, and other entities that are assumed to exist in some areas of interest and the relationships that hold among them. A conceptualization is an abstract and simplified view of the world that we wish to represent for some purpose. An ontology is an explicit specification of this conceptualization and the design of ontologies should follow a process to guide and evaluate this model. Several studies can be found, an example is the criteria identified by Gruber [6] for ontologies whose purpose is knowledge sharing and interoperation among programs based on a shared conceptualization.

There are also more detailed methodologies that seek to establish a strategy for the guided development of ontologies. In 1995 Uschold and King [31] suggested a methodology based on the following phases: identify the target ontology, build, evaluate, and document it. A more ambitious approach is the so called Methontology [32] which covers the whole life cycle, matching the ontology to a software product. Noy and McGuiness [33] for their part, describe an iterative process consisting of several stages: determine the domain and scope, reuse existing ontologies, list the important terms, define classes and their hierarchy, define properties and constraints and create the instances.

In 2010, Suarez-Figueroa [34] presented a new approach with NeOn methodology. This approach for building ontology networks is a scenario-based methodology that supports the collaborative aspects of ontology development and reuse, as well as the dynamic evolution of ontology networks in distributed environments. NeOn exposes a set of nine scenarios for building ontologies and ontology networks, emphasizing the reuse of ontological and non-ontological resources, reengineering and fusion, and considering collaboration and dynamism.

NeOn Methodology is the most suitable approach for the design of MTO, primarily due to the fact that is focused on the reuse and transformation of non-ontological resources, such as the CSV files that compose GTFS. The development is mainly ascribed to the second scenario, using adapters which allow the generation of semantic information from data sources currently available on the Internet.

### 3.2. GTFS Adapter

Given that GTFS is currently the de facto standard for transit data representation, it has been chosen to undertake a reformulation of their CSV files to entities within the ontology. A survey of existing transportation and geographical vocabularies that could be reused for the ontology design has been conducted. Thus, widely community supported vocabularies, like GeoNames [35], GeoSPARQL [36], Time [37] or WGS84 [38] have been used.

Also, an adapter, a desktop portable and multiplatform application implemented in Java, has been developed for the conversion from GTFS to MTO. Its functionality and technical characteristics are:

1. Selection and validation of GTFS files compressed in ZIP format. Extraction of the CSV files contained in the ZIP corresponding to the different concepts of a transit operation.
2. Loading of the MTO OWL file with the ontological base model. This model has been developed via OWL API, a Java API and reference implementation for creating, manipulating and serializing ontologies.
3. Transformation of non-ontological resources: establishment of entities, properties and constraints generating unique indexes for each transit agency and conversion of specific individuals from CSV files.
4. Generation of semantic appropriate links and hierarchy including links to external vocabularies like GeoNames and creation of the instances.
5. Serialization of OWL resulting files (to facilitate its treatment the number of files depends on the size of the original GTFS file).

### 3.3. MTO Specification

The pursued goal with the definition of an ontological data model is to integrate and structure existing multimodal transportation data, obtaining a semantic model for the representation of transport information. This model will have the ability to extend and link related relevant information and will be supported with tools that enable the effective consultation of its information.

As specified previously, MTO has been developed as an OWL2 ontology file, a semantic extension over RDF (XML standard for resource description) generated in 2009 by the W3C. Thanks to the use of ontologies to define the transport data model, the knowledge can be easily processed and shared between the components of the system. It also favored interoperability with other transport systems.

To facilitate the management of the ontology, it is published as two separate OWL files. Both files are needed for the operation of the ontology, so the distribution of them as a bundle is compulsory:

- *mto-core.owl* with the ontological base model; establishing the entities, properties and constraints of the ontology but without links to other vocabularies nor individuals or concrete instances.

- *mto-top.owl* with all the references generated for the entities and properties imported from other vocabularies and linked to the base ontology. *mto-core.owl* imported this file.

Protégé, a free, open-source ontology editor and framework for building intelligent systems has been used for the development of MTO. Figure 2 shows the ontology loaded in the editor.

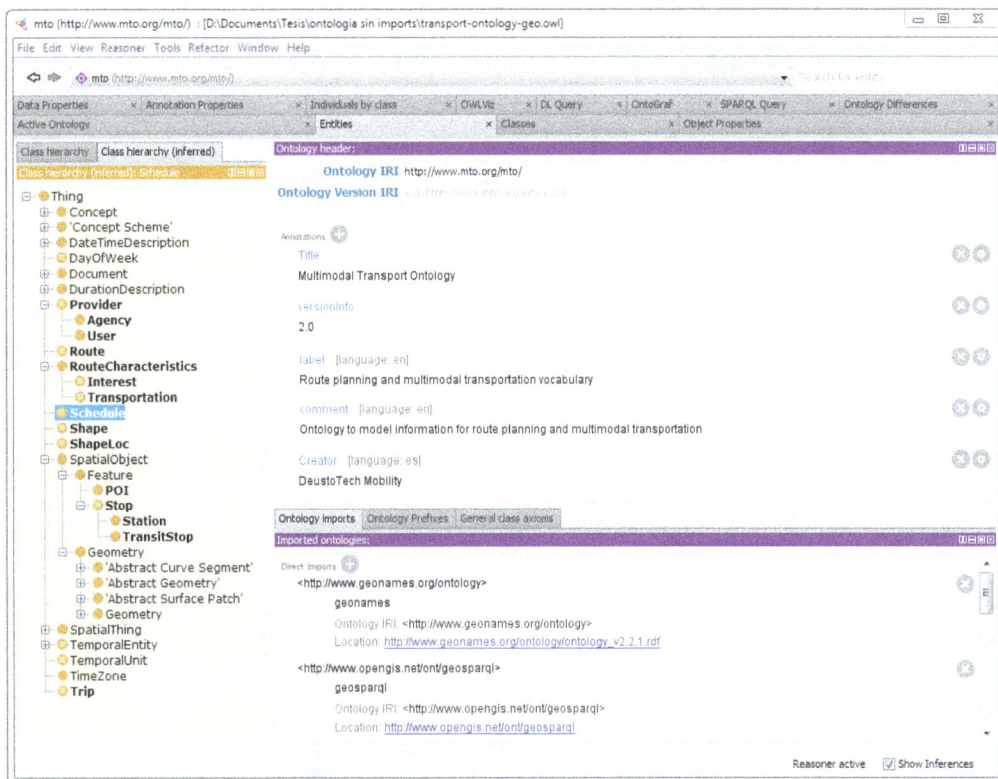

**Figure 2.** MTO OWL file in Protégé.

The ontology description can be seen in the foreground, with the title, IRI and version of the ontology, among other information. On the bottom of the UI the imported vocabularies are shown; in this case GeoNames and GeoSPARQL. The left panel in yellow shows the ontology class hierarchy, with the entities corresponding to MTO highlighted in bold. The MTO vocabulary as well as the external vocabularies imported for the generation of the transport data model will be explained in more detail in the following subsections.

### 3.3.1. MTO Vocabulary

As indicated above, the basis for the transport data model will be taken from GTFS specification. The main classes, properties and relationships of the MTO vocabulary along with a functional description of them and its correspondence in GTFS are explained below and shown in Figure 3.

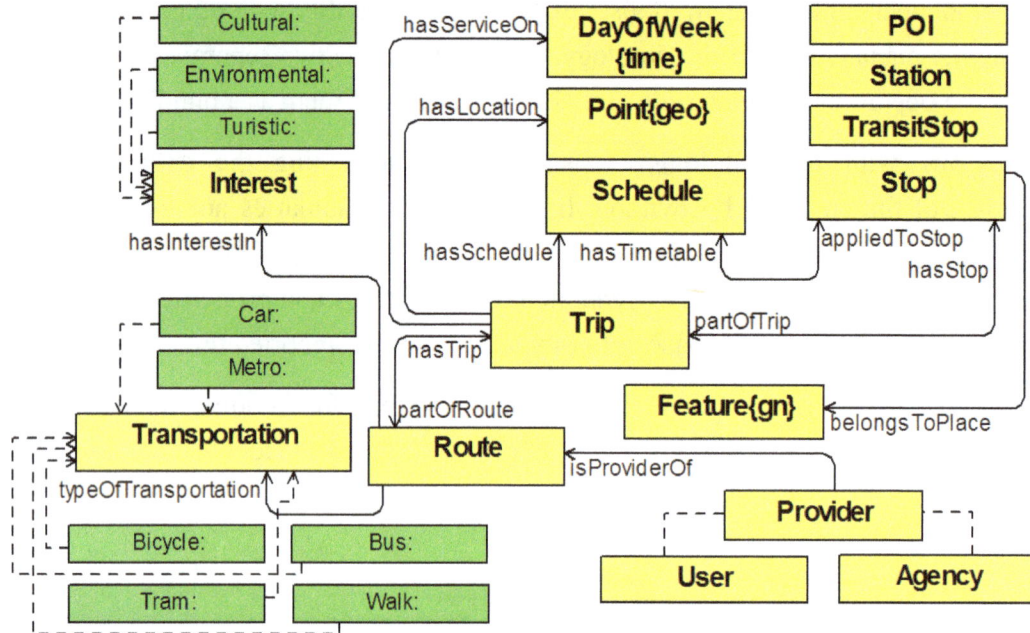

**Figure 3.** Multimodal Transport Ontology main concepts and relationships.

- *Provider*. Abstract class that will act as a parent class for all data providers. Has no equivalence in GTFS.
- *Agency*. One or more transit agencies that provide the data. Inherits from *Provider*. Equivalent to agency.txt file.
- *Route*. Transit routes. A route is a group of *Trips* that are displayed to riders as a single service. Equivalent to routes.txt file.
- *RouteCharacteristics*. Abstract class that will act as a container for the different route options defined. Has no equivalence in GTFS.
- *Transportation*. Means of transport used on a *Route*. Inherits from *RouteCharacteristics*. Has no equivalence in GTFS.
- *Trip*. Trips for each *Route*. A trip is a sequence of two or more *Stops* that occurs at specific time. Equivalent to trips.txt file.
- *Stop*. Individual locations where vehicles pick up or drop off passengers. Inherits from *Feature* (GeoSPARQL). Two significant properties: *hasGeometry* establishing a standard format to define the geometry of the particular point, and *belongsToPlace* (GeoNames) linking point position with the related geopolitical entity. Equivalent to stops.txt file.
- *TransitStop*. Specialization of *Stop* class for stops that are part of a *Trip*. Inherits from *Stop*. Has no equivalence in GTFS.
- *Schedule*. Times when a vehicle arrives at and departs from individual *Stops* for each *Trip*. Equivalent to stop_times.txt file.

- *Shape & ShapeLoc.* Rules for drawing lines on a map to represent a transit organization's routes. Equivalent to shapes.txt file.
- *POI.** Extension of the transport base ontology. Define tourist sights. Inherits from *Feature* (GeoSPARQL). Has no equivalence in GTFS.
- *POIClassification.** Extension of the transport base ontology. Define a classification for the *POIs*. Has no equivalence in GTFS.

### 3.3.2. Geographic Information Management

One of the key characteristics for a system which aims to model transportation information is the design used to store the geographical data. In this aspect, the decisions that have been taken are based on the standardization and future extensibility and/or reusability of the designed ontology.

Regarding standardization this work has followed a twofold approach: geo-referenced points have been defined by using the widely supported properties latitude and longitude, following the WGS84 standard. *hasGeometry* property has also been defined, extending from the GeoSPARQL ontology and allowing to perform geometric queries over ontology points, as shown in Figure 4.

```
SELECT DISTINCT ?poi ?lat ?lon WHERE {
  ?poi mto:hasLatitude ?lat; mto:hasLongitude ?lon.
  ?poi geo:hasGeometry ?g.?g geo:asWKT ?wkt.
FILTER (geof:distance(?wkt, "LINESTRING(-2.50 42.50,
  -2.60 42.60)"^^geo:wktLiteral, units:metre) < 5000)}
```

**Figure 4.** SPARQL query. Selecting POIs within 5 km of a given line.

As can be seen in the code snippet, the query performs a selection of POIs, including its position. To filter the resulting data, the function *geof:distance* is being used with relation to a geometric line. These kind of geometric filters are provided by GeoSPARQL, an OGC standard for geospatial semantic data. For these functions to be operational, they must be supported by the SPARQL endpoint.

```
SELECT DISTINCT ?routelongname ?nommuni ?prov WHERE {
  ?route a mto:Route; mto:hasName ?routelongname.
  ?route mto:hasTrip ?trip.
  ?trip mto:hasStop ?stop.
  ?stop mto:belongsToPlace ?muni.
  ?muni gn:name ?nommuni; gn:parentADM2 ?prov.
  ?prov gn:officialName ?nameprov

FILTER(regex(str(?nameprov),'Alava','i'))} ORDER BY ?route
```

**Figure 5.** SPARQL query. Selection of routes located in a particular province/jurisdiction.

It has also been decided to link the ontology with GeoNames, an ontology engaged in storing geopolitical information. The link is done through the *belongsToPlace* property referencing the

GeoNames URI corresponding to the territory where the point of the ontology is located. Figure 5 shows an example query using this link with GeoNames ontology.

In this case the query performs a selection of transit routes, including its name and the route location. To filter the resulting data, geopolitical information about the jurisdiction in which the route takes place has been used. The property *mto:belongsToPlace* is responsible for gather this information by linking the ontology with GeoNames data.

### 3.4. Linked Vocabularies

The representation of the transport-related information as an ontology facilitates the use of advantageous aspects, like the ability to link data with other data sources, which can be relevant to the specific domain.

The enrichment phase (Figure 1) complies with the goal of adding value to the already integrated transport data through the use of collaborative information, relying on the ability of the architecture and the proposed ontology to link these data. This supposes an improvement of the information by integrating in it non-quantitative aspects.

To do so, the ontology provides the POI class, a container for collaborative points of interest that inherits from *Feature* (GeoSPARQL) to get the geographical position (latitude and longitude) of the tourist sights. It also stores the original URI of the element, its name and description, the provider of the information and the classification assigned to the POI that has been generated according to their function: services, entertainment, tourism, *etc*.

It was decided to extend MTO by linking and classifying the collaborative points of interest provided by LinkedGeoData and GeoNames ontologies. LinkedGeoData uses the data collected by the OpenStreetMap project and makes it available as an RDF knowledge base according to the Linked Data principles. Figure 6 shows the results of a SPARQL query over MTO asking for the LinkedGeoData [39] collaborative POIs (related to restaurants) in the surroundings of Biscay, Spain.

### POI Viewer                                                                     — ✕

| POI | NOMPOI | POIPROOV | POICLASS |
|-----|--------|----------|----------|
| http://linkedgeodata.org/519238253 | Azkorri | linkedgeodata.org | mto/Hospitality |
| http://linkedgeodata.org/519238253 | Azkorri | linkedgeodata.org | mto/Bar |
| http://linkedgeodata.org/519238253 | Azkorri | linkedgeodata.org | mto/Entertainment |
| http://linkedgeodata.org/519237929 | Iturgitxi | linkedgeodata.org | mto/Hospitality |
| http://linkedgeodata.org/519237929 | Iturgitxi | linkedgeodata.org | mto/Bar |
| http://linkedgeodata.org/519237929 | Iturgitxi | linkedgeodata.org | mto/Entertainment |
| http://linkedgeodata.org/311955930 | Montecasino | linkedgeodata.org | mto/Hospitality |
| http://linkedgeodata.org/311955930 | Montecasino | linkedgeodata.org | mto/Bar |
| http://linkedgeodata.org/311955930 | Montecasino | linkedgeodata.org | mto/Entertainment |
| http://linkedgeodata.org/262455730 | Fosters | linkedgeodata.org | mto/Hospitality |
| http://linkedgeodata.org/262455730 | Fosters | linkedgeodata.org | mto/Bar |
| http://linkedgeodata.org/262455730 | Fosters | linkedgeodata.org | mto/Entertainment |

**Figure 6.** SPARQL query results. Collaborative information about restaurants in Biscay.

## 4. System Architecture

The way in which the transportation information is represented and structured supposes an innovation provided by the present work. However, the data model has to be complemented with an architecture that supports it. This is conducted by several distributed SPARQL servers. Each one of the servers maintains its own transit information and is managed in a local way, but facilitates the interoperability by means of its connection with the remaining servers by URIs, which are defined as metadata inside the ontology, allowing so, to perform distributed queries in a transparent way.

### 4.1. Distributed Transport Information

The architecture, shown in Figure 7, is similar to the one used by DNS to solve web domain names, facilitating the information distribution in a straightforward way.

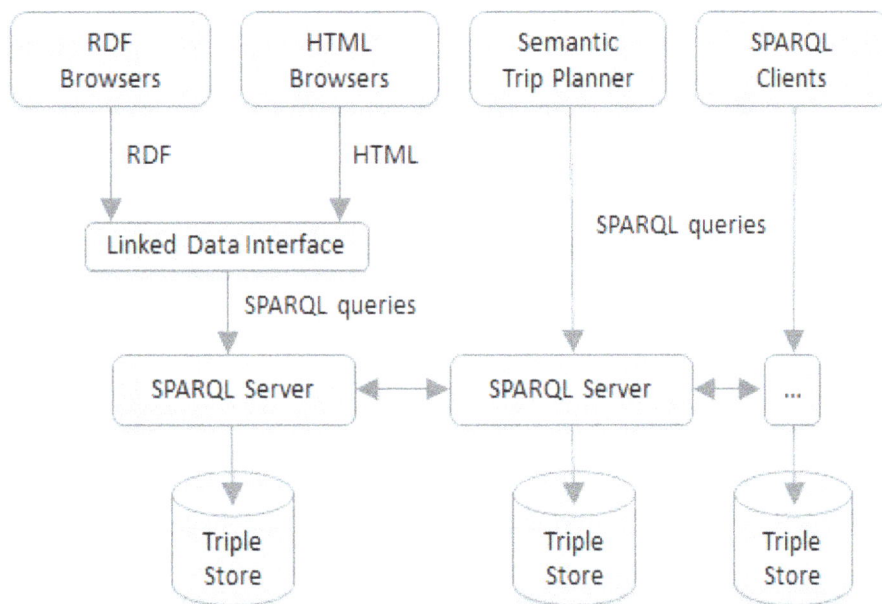

**Figure 7.** System Architecture for Multimodal Transport Information provision.

When a query reaches one of the SPARQL servers, it distributes the request to the other servers hierarchically. In this case, working with geospatial information, a geopolitical classification using administrative regions (continent, state, region, town, *etc.*) has been conducted. A SPARQL server will therefore contain information about their subsidiary administrative regions thereby achieving a complete distributed model.

The main components of the proposed distributed architecture are described below:

- *Transport Information Clients*. Different types of clients (RDF browsers, HTML browsers, SPARQL clients, *etc.*) that can request the transport information provided by the system, including the semantic trip planner developed in order to validate the architecture (further described in Section 5).

- *Linked Data Interface*. Linked Data frontend for SPARQL endpoints deployed with the aim of realizing content negotiation. It provides a data interface to RDF browsers and a simple HTML interface for HTML browsers.
- *SPARQL Servers*. Distributed set of interoperable SPARQL servers. High-performance SPARQL endpoints compatible with the GeoSPARQL standard have been used, providing so, an index for geospatial queries, making it so highly indicated in the transportation domain.
- *Triple Store*. Purpose-built database for the storage and retrieval of triples through semantic queries. A triple is a data entity composed of subject-predicate-object. Each triple store maintains its own transit information as well as links to its subsidiary SPARQL servers.

### *4.2. Software Architecture*

In the following points, the specific features and the criteria in the selection of the software tools for the development of the proposed architecture will be exposed.

### 4.2.1. Linked Data Interface: *Pubby*

*Pubby* is an RDF server that adds a Linked Data interface to existing SPARQL-capable triple stores. Many triple stores and other SPARQL endpoints can be accessed only by SPARQL client applications that use the SPARQL protocol and cannot be accessed by the growing variety of Linked Data clients. One of the goals pursued by the proposed architecture is to enable the distribution and sharing of multimodal transport data. This objective would not be properly satisfied if the information were only accesible via SPARQL queries or using semantic applications. Figure 8 shows the deployment of *Pubby* for MTO information provision.

In RDF, resources are identified by URIs. *Pubby* will handle requests to the mapped URIs by connecting to the SPARQL endpoint, asking it for information about the original URI, and passing back the results to the client. It also handles various details of the HTTP interaction and content negotiation between HTML, RDF/XML and Turtle descriptions of the same resource.

**Figure 8.** Deployment of *Pubby* for MTO information provision.

### 4.2.2. SPARQL Server: *Parliament*

*Parliament* is a high-performance triple store designed for the Semantic Web and compatible with the RDF, RDFS, OWL, SPARQL, and GeoSPARQL standards. It was released as an open source project under the BSD license in June, 2009. It offers a number of interesting features highly indicated for the transportation scenario:

- Employs an innovative data storage scheme that interweaves the data with a unique index. Because of that, it can answer queries efficiently by reordering query execution so that the most restrictive parts are executed first [40]. This is an important feature for the proposed solution due to the execution of complex queries related to geospatial and contextual data.

- Has a temporal index, so that it can efficiently answer queries related to time intervals.

- Supports GeoSPARQL, the newly adopted OGC standard for geospatial semantic data. Using its geospatial index, it can efficiently answer queries like "find items located within region X".

- Includes a high-performance rule engine. This enables *Parliament* to automatically and transparently infer additional facts and relationships in the data to enrich query results. It implements RDFS inference plus selected elements of OWL (equivalent classes and properties, and inverse, symmetric, functional, inverse functional and transitive properties).

The architecture and characteristics described, as well as the support for SPARQL 1.1 Service description and Federated queries, made *Parliament* an optimal option for the storage and management of the transportation data that has to be integrated and shared by the platform.

### 4.3. Federated SPARQL Queries

SPARQL can be used to express queries across diverse data sources, whether the data is stored natively as RDF or viewed as RDF via middleware. The Federated Query specification [41] defines the syntax and semantics for executing queries distributed over different SPARQL endpoints. The SERVICE keyword extends SPARQL 1.1 to support queries that merge data distributed across the Web.

The developed architecture for semantic transport information provision makes extensive use of this specification to support the aggregation of distributed information related to multimodal mobility.

To do so, a set of resources (*Datasets*) will be described in the ontology as references to other hierarchically classified SPARQL servers. In the following code fragment (Figure 9) the description of a *Dataset* for the *Araba* region, including the SPARQL endpoint URL, is established.

```
<rdf:RDF xmlns:mtof="http://www.mto.org/fed/">
   <rdf:Description rdf:about="mtof:Dataset/araba">
   <mtof:endpoint rdf:resource="http://localhost:3030/sparql"/>
   <mtof:webview rdf:resource="http://localhost:8081"/>
   </rdf:Description>

</rdf:RDF>
```

**Figure 9.** Dataset resource description, including the SPARQL endpoint URL.

Once the references have been set, the queries to the server will be distributed accordingly, and the information related to one specific area will be gathered in a distributed and transparent way.

Figure 10 shows an example of a federated query that receives all the information about *Stops* contained in the associated SPARQL servers. This is done using the variable *?ep* that represents a *mtof:endpoint* like the one described in Figure 9. The server iterates over all the endpoints and, using the SERVICE protocol, performs remote queries to all of them, aggregating their data and returning the information to the original queried server.

```
PREFIX rdf: <http://www.w3.org/1999/02/22-rdf-syntax-ns#>
PREFIX mto: <http://www.mto.org/mto/>
PREFIX mtof: <http://www.mto.org/fed/>

SELECT ?stopid ?stopname ?stoplat ?stoplon ?muni ?n
WHERE {
    ?ds mtof:endpoint ?ep.
SERVICE ?ep {
    ?stopid a mto:Stop.
    ?stopid mto:hasName ?stopname.
    ?stopid mto:hasLatitude ?stoplat.
    ?stopid mto:hasLongitude ?stoplon. }
} ORDER BY ?stopid
```

**Figure 10.** Federated SPARQL query.

## 5. STP: Semantic Trip Planner

As pointed out in the state of the art, multimodal trip planners are one of the software tools that have advanced more in recent years within the field of ITS, providing to the user information about itineraries, timetables, routing and other transit data for an intermodal passenger transport journey.

The implementation of a semantic trip planner, an extension over the traditional journey planner, based on the integration of Linked Open Data together with collaborative information and running on the proposed architecture was established in order to validate the developed solution. The proposed Sematic Trip Planner will make use of the distributed architecture and the ontology-based data model to give users relevant information related to multimodal transportation in a more convenient way.

For the implementation of STP and the later tests we used a 64-bit machine with a quad-core 2.66 GHz processor and 8 GB of RAM. The machine was running Windows 7 and Java 7 64-bits. To deploy all the components of the architecture (Parliament, Pubby and STP) we recommend a machine with at least 4 GB of RAM with no restrictions regarding platform or OS.

*5.1. OpenTripPlanner*

The goal of the proposed solution is to provide semanticized transport information to the user in a transparent way. A multimodal trip planner with a map-based web interface is thus a good platform for testing the achievement of this goal. OpenTripPlanner (OTP), an open source platform for multimodal journey planning has been selected as a base platform for the STP software solution development.

OTP follows a client-server model, providing several map-based web interfaces (Figure 11) as well as a REST API to be used by third-party applications. It relies on open data standards including GTFS for transit and OpenStreetMap (OSM) for street networks.

Launched in 2009, the project has attracted a thriving community of users and developers, receiving support from public agencies, startups, and transportation consultancies alike. There are multiple OTP deployments and is also the routing engine behind several popular smartphone applications.

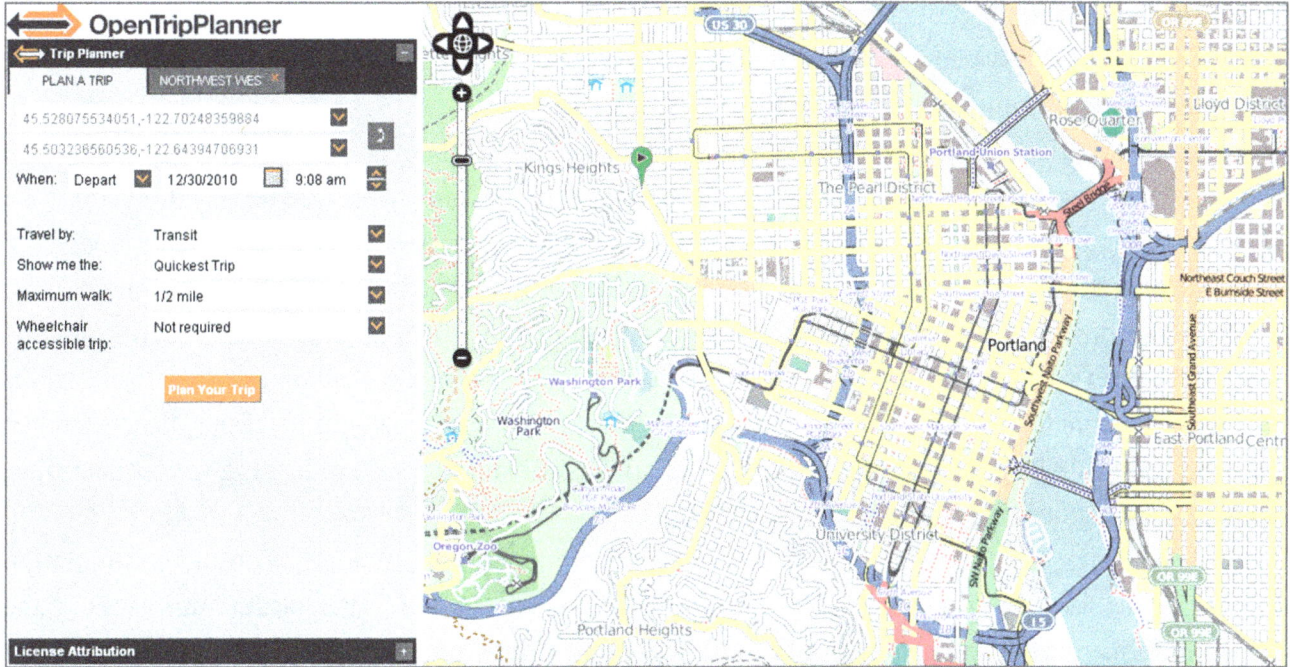

**Figure 11.** OpenTripPlanner map-based web interface.

The solution presented will use the features provided by OTP, such as the map-based web interface and the routing engine. This in turn leads to an extensive modification of the OTP project (possible thanks to its collaborative and open source nature) to take, for example, the available SPARQL servers as a data source for the transport information, instead of the GTFS files.

## 5.2. STP Functionality and Improvements

Besides the deployment and initial configuration of the OTP project, which is outside the scope of this article, the following actions have been undertaken for the implementation of STP:

1. Load of transport data published by the Basque Government and generation of the corresponding semantic content via the developed GTFS adapter (detailed in Section 3.2).
2. Implementation of three distributed SPARQL servers (detailed in Section 4.1), publishing as LOD semanticized transit information relating to the three Basque country provinces.
3. OTP project modifications to include data provided by the supplied architecture:

   o Consuming the ontological model developed
   o Accessing existing related information
   o Providing contextualized transit information to the user

As can be seen, the last point includes several changes over OTP related with the semantic nature of the transport information now provided. These modifications will be properly described in the next subsections.

## 5.2.1. Data Sources Configuration

OTP works with a structure called *Graph* that contains all the information about the topology of the area in which the routing algorithm has to operate. This *Graph* is generated according to the transit data provided to the system. To provide this data, an xml file has to be submitted with a property pointing to the path with the GTFS files the system will consume.

This operative has some flaws that have been satisfactorily resolved with the use of distributed SPARQL servers for storing the transit information. On the one hand, the GTFS files must be downloaded manually from each of the transit information providers (agencies, administrations, *etc.*). It should also be noted that such information must be regularly updated to maintain its accuracy.

With the deployment of STP, the xml file only stores an URL to a SPARQL server that, as mentioned previously, contains a hierarchical classification with the rest of servers. Each server maintains information about local agencies, so the update of its contents will be easier, faster and can be done via a SPARQL update query. So there is no need of storing or downloading transit data files, and the information will always remain updated.

Some changes have been made in OTP to support this operative. In addition to the configuration files, a new class inheriting from *GraphBuilder* has been created. This class controls how the transit information is processed for the building of the final topology graph. A solution for the transformation of RDF data (coming from the SPARQL servers) to CSV data, which is needed by OTP, has been implemented.

Thanks to the flexibility of SPARQL and *Parliament*, a set of standard queries have been designed that, starting from the distributed semantic information contained on the triple stores, generate CSV formatted responses with all the needed information, directly and transparently. The code snippet shown in Figure 10 is an example of this type of query, in this case receiving data about the stops on route and formatting the output in CSV.

## 5.2.2. Context Management and LOD

An important topic to consider when generating more accurate and rich information is the specific context of the user. Without this context, data offered could lose interest, since no personalized information can be generated. For each query that arrives to STP, the user context is established with the purpose of filtering data according to their specific circumstances. The architecture defines, in order to build such model, a faceted search with a set of properties referred to the POIs in the route:

- Name of the point of interest to look for.
- Linked Data source to consult (GeoNames and/or LinkedGeoData).
- Geographical location of the query. This information can be used:

  o In a geopolitical way (e.g., POIs located in Biscay).
  o In a geometric way (e.g., POIs within 5 km of a given route).

- Hierarchical classification of the POIs

    o Services: Transport, Accommodation, Utilities
    o Entertainment: Facilities, Stores, Hospitality
    o Point of Interest: Cultural, Environmental, Touristic

An SPARQL query is dynamically generated based on the selections made by the user. The query can combine all of the parameters listed previously, like the geographical position of the user, the name of the POI or its classification, the proximity to the optimal route, *etc.*

The query is generated transparently to the user, who simply makes use of the different options provided for that purpose in the web interface (Figure 12). Internally, the tool uses the geographic data management capabilities and the integration of relevant information to adequately answer the requests.

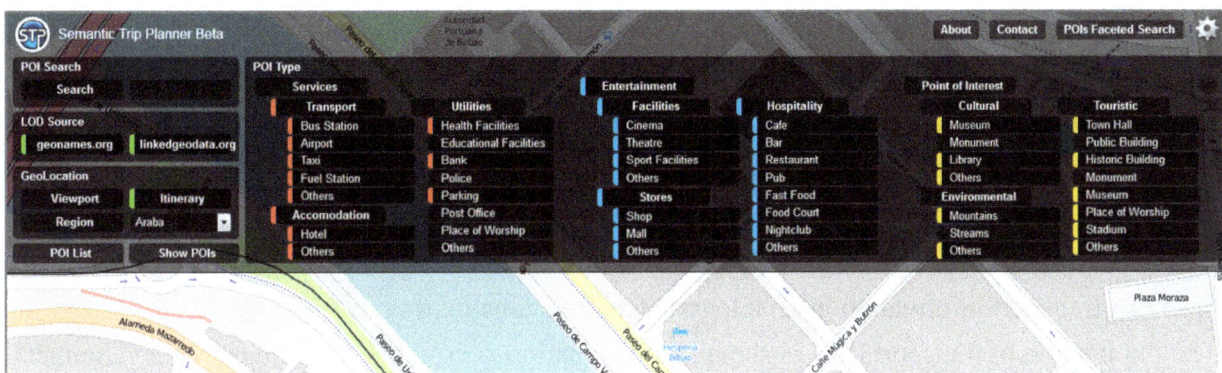

**Figure 12.** STP. Faceted search for the selection of POIs.

The generated query, resulting from the selection made by the user, could include related information coming from data sources or services that have been integrated in the ontology. In this way, the use of context and LOD can tailor information to specific user needs, customizing and improving the information given and providing relevance to it.

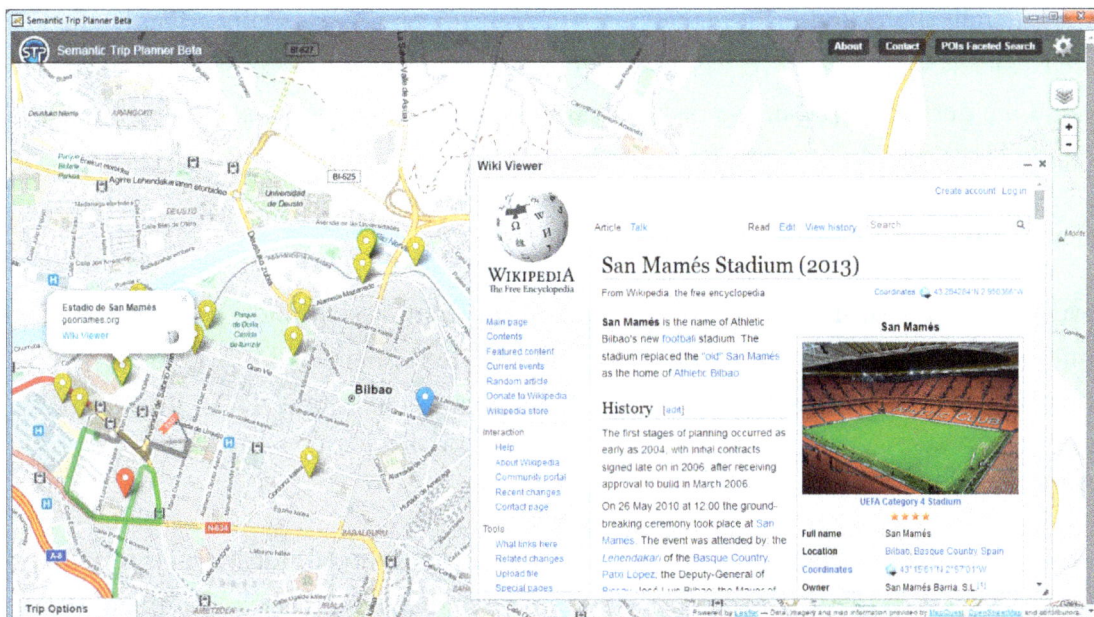

**Figure 13.** STP. Extended information (Wikipedia) about POIs.

Figure 13 shows, for example, information from Wikipedia associated with one of the POIs. This is due to the link between MTO and GeoNames, the ontology that provides such information.

## 6. Conclusions

The study conducted on the systems currently used to represent transport information led us to address the challenge of the management and representation of transport information, often based on the use of heterogeneous databases or closed non-interoperable formats that were not sufficiently interoperable, extensible or open. However, it has also shown how new trends in the management of transit data, as the generation of ontologies for representing geospatial data or the use of the Web of Data to link standard geographic information, applied to the transportation domain, can help in the task of structuring, distributing and sharing currently available information.

The work highlighted in this paper presents an alternative solution, a new approach to the management and exchange of transit data with the development of a novel software architecture, with particular emphasis on the design of the data model and the enablement of services for information retrieval, thereby providing semantic transit information, formally structured and integrated. With that, and following the strategy route pointed out by the European Commission white paper on transport [42], it is intended to improve and extend the public institutions and companies' capability to obtain, share, and provide transport information in an integrated and sustainable way.

The publication of transport data as semantic information is established through the development of an ontology for multimodal transportation (MTO) and the design of an architecture of distributed and interoperable SPARQL servers. A much more powerful and direct tool for accessing to all the available transit data, with support for advanced queries (geospatial and geopolitical), multiple output formats and enabling dynamic integration of information.

The work done, based on the application of the Linked Open Data principles to the field of transportation has tried to improve the quality of the provided information. This promotes the development of applications that could provide innovative services in an area, sustainable mobility, which are receiving social and institutional support given its environmentally friendly characteristics. The Linked Data community is very active and hence the next steps point to an extension of the ontology and the desirable appearance of an ecosystem of both applications and open transit data that supports it.

## Acknowledgments

This work has been partially funded by the European Commission under the Seventh Framework Programme (Grant agreement No.: 317671) and by the Spanish Ministerio de Economía y Competitividad under the "Retos de la Sociedad" funding program (Grants RTC-2014-3061-4).

## Author Contributions

Asier Moreno and Asier Perallos are responsible for the multimodal transport information software architecture design and development. Asier Moreno and Diego Lopez de Ipiña designed the multimodal transport ontology, establishing also the SPARQL queries and the integration of relevant

LOD. Enrique Onieva and Antonio Masegosa conducted the characterization of the context management and participate in the design of the faceted search for STP. Itziar Salaberria participates in the design of the semantic trip planner and the management of geospatial information for MTO. Asier Moreno and Itziar Salaberria prepared the manuscript.

**Conflicts of Interest**

The authors declare no conflict of interest.

**References**

1. International Association of Public Transport (UITP). Doubling the market share of public transport worldwide by 2025. Available online: http://www.slocat.net/doubling-market-share-public-transport-worldwide-2025 (accessed on 20 May 2015).
2. Weiland, R.J.; Purser, L.B. Intelligent transportation systems. In *Transportation in the New Millennium*; Transportation Research Board: Washington, DC, USA, 2000.
3. Boriboonsomsin, K.; Barth, M.J.; Zhu, W.; Vu, A. Eco-routing navigation system based on multisource historical and real-time traffic information. *IEEE Trans. Intell. Transp. Syst.* **2012**, *13*, 1694–1704.
4. Barfield, W.; Dingus, T.A. *Human Factors in Intelligent Transportation Systems*; Psychology Press: New York, NY, USA, 2014.
5. Campbell, J.L.; Carne, C.; Kantowitz, B.H. Human Factors Design Guidelines for Advanced Traveler Information Systems (ATIS) and Commercial Vehicle Operations (CVO). In *Technical Report FHWA-RD-98–057*; U.S. Department of Transportation: Seattle, WA, USA, 1998.
6. Gruber, T.R. Toward principles for the design of ontologies used for knowledge sharing. *Int. J. Hum. Comput. Study* **1995**, *43*, 907–928.
7. Fensel, D. Ontologies. In *Ontologies*; Springer: Berlin/Heidelberg, Germany, 2001; pp. 11–18.
8. Gruber, T.R. A translation approach to portable ontology specifications. *Knowl. Acquis.* **1993**, *5*, 199–220.
9. Malik, Z.; Rezgui, A.; Medjahed, B.; Ouzzani, M.; Sinha, A.K. Semantic integration in geosciences. *Int. J. Semant. Comput.* **2010**, *4*, 301–330.
10. Berners-Lee, T.; Fischetti, M.; Dertouzos, M.L. *Weaving the Web: The Original Design and Ultimate Destiny of the World Wide Web by Its Inventor*; HarperInformation: New York, NY, USA, 2000.
11. Berners-Lee, T.; Bizer, C.; Heath, T. Linked Data the story so far. *Int. J. Semant. Web Inf. Syst.* **2009**, *5*, 1–22.
12. Berners-Lee, T. Is your Linked Open Data 5 Star? Available online: http://www.w3.org/DesignIssues/LinkedData.html (accessed on 5 May 2015).
13. Fensel, D.; Facca, F.M.; Simperl, E.; Toma, I. *Semantic Web Services*; Springer: Berlin, Germany, 2011.
14. Gene Ontology Consortium. Gene Ontology annotations and resources. *Nucleic Acids Res.* **2013**, *41*, D530–D535.

15. Mol, A. *The Body Multiple: Ontology in Medical Practice*; Duke University Press: Durham, NC, USA, 2002.

16. Dascalu, M.I.; Bodea, C.N.; Lytras, M.; de Pablos, P.O.; Burlacu, A. Improving e-learning communities through optimal composition of multidisciplinary learning groups. *Comput. Hum. Behav.* **2014**, *30*, 362–371.

17. Jia, H.; Wang, M.; Ran, W.; Yang, S.; Liao, J.; Chiu, D. Design of a performance oriented workplace e-learning system using ontology. *Expert Syst. Appl.* **2011**, *38*, 3372–3382.

18. Dong, H.; Hussain F.K.; Chang, E. Transport service ontology and its application in the field of semantic search. In Proceedings of IEEE International Conference on Service Operations and Logistics, and Informatics, 2008 (IEEE/SOLI 2008), Beijing, China, 12–15 October 2008; Volume 1, pp. 820–824.

19. De Oliveira, K.M.; Bacha, F.; Mnasser, H.; Abed, M. Transportation ontology definition and application for the content personalization of user interfaces. *Expert Syst. Appl.* **2013**, *40*, 3145–3159.

20. Boucher, S.; Zimanyi, E. An ontology-based geodatabase interoperability platform. In *Cases on Semantic Interoperability for Information Systems Integration: Practices and Applications*; IGI Global: Hershey, PA, USA; pp. 294–315.

21. Janowicz, K. Observation-driven geo-ontology engineering. *Trans. GIS* **2012**, *16*, 351–374.

22. Lorenz, B.; Ohlbach, H.J.; Yang, L. Ontology of Transportation Networks. Available online: http://rewerse.net/deliverables/m18/a1-d4.pdf (accessed on 20 May 2015).

23. Lemmens, R.L.G. Semantic Interoperability of Distributed Geoservices. PhD Thesis, Delft University of Technology, Delft, The Netherlands, 2006.

24. Zhao, T.; Zhang, C.; Wei, M.; Peng, Z.R. Ontology-Based Geospatial Data Query and Integration. In *Geographic Information Science of Lecture Notes in Computer Science*; Springer: Berlin, Germany, 2008; pp. 370–392.

25. Niaraki, A.S.; Kim, K. Ontology based personalized route planning system using a multi-criteria decision making approach. *Expert Syst. Appl.* **2009**, *36*, 2250–2259.

26. Houda, M.; Khemaja, M.; Oliveira, K.; Abed, M. A public transportation ontology to support user travel planning. In Proceedings of 2010 Fourth International Conference on Research Challenges in Information Science (RCIS), Nice, France, 19–21 May 2010.

27. Gunay, A.; Akcay, O.; Altan, M.O. Building a semantic based public transportation geoportal compliant with the INSPIRE transport network data theme. *Earth Sci. Inf.* **2014**, *7*, 25–37.

28. Dey, A.K.; Abowd, G. The context toolkit: Aiding the development of context-aware applications. In Proceedings of Workshop on Software Engineering for Wearable and Pervasive Computing, Limerick, Ireland, 6 June 2000.

29. Kim, J.; Chung, K.Y. Ontology-based healthcare context information model to implement ubiquitous environment. *Multimed. Tools Appl.* **2014**, *71*, 873–888.

30. Buján, D.; Martín, D.; Torices, O.; López-de Ipiña, D.; Lamsfus, C.; Abaitua, J.; Alzua-Sorzabal, A. Context Management Platform for Tourism Applications. *Sensors* **2013**, *13*, 8060–8078.

31. Uschold, M.; King, M. Towards a Methodology for Building Ontologies. In *Workshop on Basic Ontological Issues in Knowledge Sharing*; The University of Edinburgh: Edinburgh, UK, 1995.

32. Fernández, M.; Gómez, A.; Juristo, N. Methontology: From ontological art towards ontological engineering. In *Proceedings of the Ontological Engineering AAAI-97 Spring Symposium Series*; American Asociation for Artificial Intelligence: Stanford, CA, USA, 1997.

33. Noy, N.F.; McGuinness, D.L. Ontology Development 101: A Guide to Creating Your First Ontology. Available online: http://protege.stanford.edu/publications/ontology_development/ ontology101.pdf (accessed on 20 May 2015).

34. Suárez-Figueroa, M.C. NeOn Methodology for Building Ontology Networks: Specification, Scheduling and Reuse. Ph.D. Thesis, Universidad Politécnica de Madrid, Madrid, Spain, 2010.

35. GeoNames Ontology. Available online: http://www.geonames.org/ontology (accessed on 5 May 2015).

36. GeoSPARQL—A Geographic Query Language for RDF Data. Available online: http://www.opengeospatial.org/standards/geosparql (accessed on 5 May 2015).

37. Time Ontology in OWL. Available online: http://www.w3.org/tr/owl-time (accessed on 5 May 2015).

38. Basic Geo (WGS84 lat/long) Vocabulary. Available online: http://www.w3.org/2003/01/geo (accessed on 5 May 2015).

39. LinkedGeoData. Available online: http://linkedgeodata.org (accessed on 5 May 2015).

40. Kolas, D.; Emmons, I.; Dean, M. Efficient Linked-List RDF Indexing in Parliament. In Proceedings of the 5th International Workshop on Scalable Semantic Web Knowledge Base Systems (SSWS), Washington, DC, USA, 26 October 2009.

41. SPARQL 1.1 Federated Query. Available online: http://www.w3.org/tr/sparql11-federated-query (accessed on 5 May 2015).

42. European Commission. *White Paper on transport: Roadmap to a single European Transport Area*; Publications Office of the European Union: Luxembourg, Luxembourg, 2011.

# Characterization of Wheat Varieties Using Terahertz Time-Domain Spectroscopy

**Hongyi Ge [1,2,*], Yuying Jiang [1,2], Feiyu Lian [3], Yuan Zhang [3] and Shanhong Xia [1,2]**

[1]  State Key Laboratory of Transducer Technology, Institute of Electronics,
    Chinese Academy of Sciences, Beijing 100080, China; E-Mails: jiangyuying11@163.com (Y.J.);
    shxia@mail.ie.ac.cn (S.X.)
[2]  University of the Chinese Academy of Sciences, Beijing 100080, China
[3]  Key Laboratory of Grain Information Processing & Control, Ministry of Education,
    Zhengzhou 450001, China; E-Mails: lfywork@163.com (F.L.); zhangyuan@haut.edu.cn (Y.Z.)

*  Author to whom correspondence should be addressed; E-Mail: gehongyi2004@163.com

Academic Editor: W. Rudolf Seitz

**Abstract:** Terahertz (THz) spectroscopy and multivariate data analysis were explored to discriminate eight wheat varieties. The absorption spectra were measured using THz time-domain spectroscopy from 0.2 to 2.0 THz. Using partial least squares (PLS), a regression model for discriminating wheat varieties was developed. The coefficient of correlation in cross validation (R) and root-mean-square error of cross validation (RMSECV) were 0.985 and 1.162, respectively. In addition, interval PLS was applied to optimize the models by selecting the most appropriate regions in the spectra, improving the prediction accuracy ($R = 0.992$ and RMSECV = 0.967). Results demonstrate that THz spectroscopy combined with multivariate analysis can provide rapid, nondestructive discrimination of wheat varieties.

**Keywords:** terahertz time-domain spectroscopy; wheat varieties; absorption spectrum; interval partial least squares

## 1. Introduction

Wheat is one of the most important agricultural products in the world. Due to increasing free trade, wheat varieties from diverse origins are widely available in the markets. However, the nutrition and processing quality of wheat varieties differ, and false wheat seeds can cause great losses for farmers. Thus, there is a demand for rapid analytical methods to discriminate the material properties of wheat. Traditional methods such as morphology analysis, physics, chemical methods, machine vision, and DNA molecular marker analysis enable sensitive classification of wheat varieties [1–4]. However, these methods are time-consuming and require complex operation. Spectroscopic methods such as near infrared, mid-infrared, and Raman spectroscopy are adequate analytical tools that have been used largely for material classification [5–7]. However, few studies have utilized the far infrared or terahertz (THz) band for discriminating wheat varieties.

Terahertz radiation, which occupies frequencies between 0.1 and 10 THz, lies between microwaves and infrared bands in the electromagnetic spectrum. Due to the rapid advances in generating, detecting, and analyzing THz radiation, THz spectroscopy has been recently developed as an analytical method [8,9]. Because many molecules have unique spectral fingerprints in THz band as a result of the vibrational transitions of the molecules, THz spectroscopy can be used for discriminating materials. Furthermore, THz radiation is non-destructive and can penetrate many nonpolar materials. As a result, THz spectroscopy has been successfully applied in detecting explosives [10] and drugs [11], in the field of biological sciences [12], and in food safety control [13].

The aim of the present study is to investigate the potential of THz spectroscopy as a non-destructive method to discriminate wheat varieties. The THz spectra of wheat varieties were measured and analyzed in the frequency of 0.2–2 THz. In addition, chemometric methods were used to evaluate wheat varieties based on THz spectra. The partial least squares (PLS) and interval PLS (iPLS) methods were used to obtain better discrimination results.

## 2. Materials and Methods

### 2.1. Experimental Setup

A conventional terahertz transmission spectroscopy system was used in the experiment. The mode-locked Ti-sapphire femtosecond laser, which provided 100-fs pulses at a wavelength of 800 nm and a repeating frequency of 80 MHz, was divided into two beams (pump beam and probe beam) using a polarization beam splitter (PBS). The THz pulses were generated from the low-temperature-grown GaAs photoconductive antenna with an attached silicon hyperhemispherical lens. The THz radiation from the emitter was collected and focused on the sample by a pair of parabolic mirrors (PM). Electro-optic (EO) detection was employed to observe the THz signal. The transmitted THz radiation was focused and collimated by PM onto the ZnTe EO detector crystal. A detailed description of THz-TDS can be found in Ref. [14]. The THz beam path was filled with nitrogen gas to remove absorption of atmospheric water vapor [15]. The samples were placed at the focal point of the THz beam spectroscope, and the measurements were performed at an ambient temperature of 294 K with a relative humidity of approximately 3%.

Using THz-TDS, we can measure both the phase and amplitude of the THz pulses propagating through the sample and reference (nitrogen gas). A reference pulse signal, $E_{ref}(t)$, in the absence of wheat and a sample pulse signal, $E_{sam}(t)$, are recorded. Comparing the sample pulse and reference pulse using a fast Fourier transform, the complex refractive index $N(\omega)$ can be expressed as follows:

$$N(\omega) = n(\omega) - ik(\omega) \tag{1}$$

where $n(\omega)$ and $k(\omega)$ are the real refractive index and extinction coefficient, respectively, describing the dispersion and absorption characteristics of the sample. Here, $\omega$ is the cyclic frequency, and $i$ is the imaginary unit. The complex transmittance function $H(\omega)$ of sample is given by [16–18]:

$$H(\omega) = \frac{E_{sam}(\omega)}{E_{ref}(\omega)} = \frac{4N}{(N+1)^2} \exp(\frac{i\omega(N-1)d}{c}) = \rho(\omega)\exp(i\phi(\omega)) \tag{2}$$

where $E_{sam}(\omega)$ and $E_{ref}(\omega)$ are the complex amplitudes of the Fourier transform of $E_{ref}(t)$ and $E_{sam}(t)$, respectively, $c$ is the speed of light, and $\rho(\omega)$ and $\phi(\omega)$ are the amplitude ratio and related phase difference of the reference and sample, respectively.

The $n(\omega)$ and absorption coefficient $a(\omega)$ were obtained by:

$$n(\omega) = \frac{\phi(\omega)}{\omega d}c + 1 \tag{3}$$

$$a(\omega) = \frac{2}{d}\ln\left(\frac{4n(\omega)}{\rho(\omega)[n(\omega)+1]^2}\right) \tag{4}$$

### 2.2. Sample Preparation

In this study, eight wheat varieties were prepared for the analysis. Samples were supplied by Henan University of Technology, Zhengzhou, China, and were harvested in 2013. The collection of samples is as diverse as possible and representative of the main production areas. These wheat grains are mixtures with different components and complex structures and can have different chemical and physical properties. The sample properties are shown in Table 1.

**Table 1.** The Properties of the eight wheat varieties under consideration.

| No. | Wheat Variety | Bulk Density (g/L) | Crude Protein Content (%) | Water Content (%) | Imperfect Grain (%) | Gluten Content (%) |
|-----|---------------|--------------------|---------------------------|-------------------|---------------------|--------------------|
| 1 | Zhengmai 9023 | 756 | 14.6 | 12.5 | 3.2 | 27.9 |
| 2 | Zhouyuan 9369 | 790 | 14.9 | 12.5 | 3.4 | 33.0 |
| 3 | Aobiao | 845 | 14.0 | 11.5 | 1.0 | 26.0 |
| 4 | DNS | 840 | 14.5 | 11.9 | 1.6 | 38.0 |
| 5 | Jiamai | 830 | 13.8 | 12.2 | 1.8 | 39.0 |
| 6 | Jinan 17 wheat | 773 | 15.6 | 11.5 | 3.0 | 34.0 |
| 7 | Zhoumai 27 | 798 | 13.2 | 12.1 | 3.8 | 33.0 |
| 8 | Yunong 416 | 787 | 14.3 | 12.5 | 4.0 | 32.5 |

The eight wheat samples are identified as Zhengmai 9023, Zhouyuan 9369, Aobiao, DNS, Jiamai, Jinan17 wheat, Zhoumai 27, and Yunong 416 wheat. For each variety of wheat, 20 samples were prepared without further purification before grinding. Generally, the wheat samples can transform to more stable form during the storage and manufacture process, such as grinding and compaction. To form thin, circular slices samples, wheat samples were ground into fine powder for 2 min, which was subsequently sieved by filtering laws using 200-eye sieves; then, the sieved powder was pressed into pellets with a thickness of approximately 1 mm and a diameter of 13 mm under a pressure of 5 tons for 5 min. All sample preparation processes were implemented at room temperature.

## 2.3. Chemometrics Methods

A chemometric analysis was performed to investigate the THz spectral data of wheat varieties using a partial least squares (PLS) regression. The PLS regression is based on latent variables, which are constructed to identify the maximal covariance between two matrices [19]. In particular, PLS was applied to find the best correlation between the spectral data $X$ and measured parameter designating the class of interest $Y$. The model is formulated as follows:

$$X = TP^{\mathrm{T}} + E \tag{5}$$

$$Y = UQ^{\mathrm{T}} + F \tag{6}$$

where $X$ and $Y$ are the input and output matrices, $T$ and $U$ are the score matrices, $P$ and $Q$ are the loading matrices that can be regarded as the covariance between $X$ and $Y$ and between $Y$ and $U$, respectively, and $E$ and $F$ are the residual matrices. In this paper, the absorption coefficient and refractive index of a selected frequency range are used as the input matrix $X$, while the wheat varieties of samples are used as the output matrix $Y$, which can be regarded as the number for each wheat variety sample, and the corresponding wheat variety can be described using number 1–8 given in Table 1.

The interval partial least squares (iPLS) method is a variable selection technique [20] used for identifying the important spectral regions and removing interference from other regions. The iPLS method divides the whole spectrum into subintervals of equal width and develops the PLS model for each subinterval spectrum. The prediction performance of PLS and iPLS is compared by the root-mean-square error of cross-validation (RMSECV) values, and the best subinterval can be selected by the lowest model RMSECV [21]. The quality of the calibration model is evaluated using the correlation coefficient between the reference and predicted value (R), the root-mean-square error (RMSE) of the calibration set (RMSEC), and the root mean square error of the prediction set (RMSEP) [22]. A better model has a better prediction accuracy that provides higher R and lower RMSECV values. The parameters are calculated as follows:

$$RMSECV = \sqrt{\frac{1}{n}\sum_{i=1}^{n}(y_r^{\ i} - y_p^{\ i})^2} \tag{7}$$

$$R = \frac{\sum_{i=1}^{n}(y_r^{\ i} - \overline{y_r})(y_p^{\ i} - \overline{y_p})}{\sqrt{\sum_{i=1}^{n}(y_r^{\ i} - \overline{y_r})^2 \sum_{i=1}^{n}(y_p^{\ i} - \overline{y_p})^2}} \tag{8}$$

where $n$ is the number of samples in the calibration sample set, $y_r^i$ is the reference value of the ith sample, $y_p^i$ is the predicted value of the ith sample, $\overline{y_r}$ is the average of the sample reference values, and $\overline{y_p}$ is the average of the predicted values of samples.

## 3. Results and Discussion

### 3.1. Spectra of Wheat Samples

To remove the random error and increase the signal-to-noise ratio (SNR), each sample is measured three times; the sample spectrum was the average of three scanning spectra in the range of 0.2–2.5 THz, and the reference was measured between every three samples. Figure 1a,b show the time-domain spectra of the eight wheat samples and the corresponding frequency-domain spectra obtained using a fast Fourier transform algorithm. Furthermore, the refractive indices and the absorption coefficients of the eight wheat samples are calculated using Equations (3) and (4), respectively, and the calculated results are shown in Figure 2a,b.

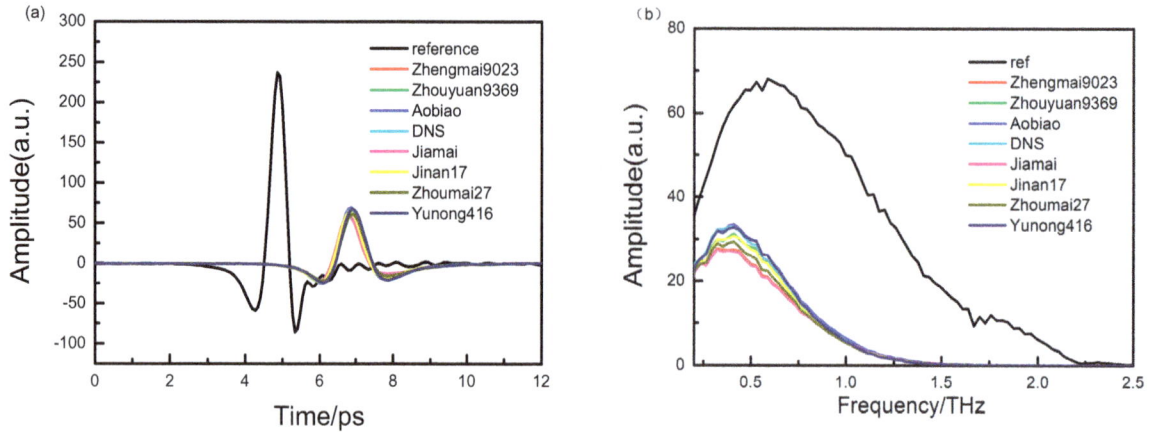

**Figure 1. (a)** Time-domain THz spectra of the eight wheat samples and reference; and **(b)** the frequency spectra of the eight wheat samples and the reference in the range of 0.2–2.5 THz.

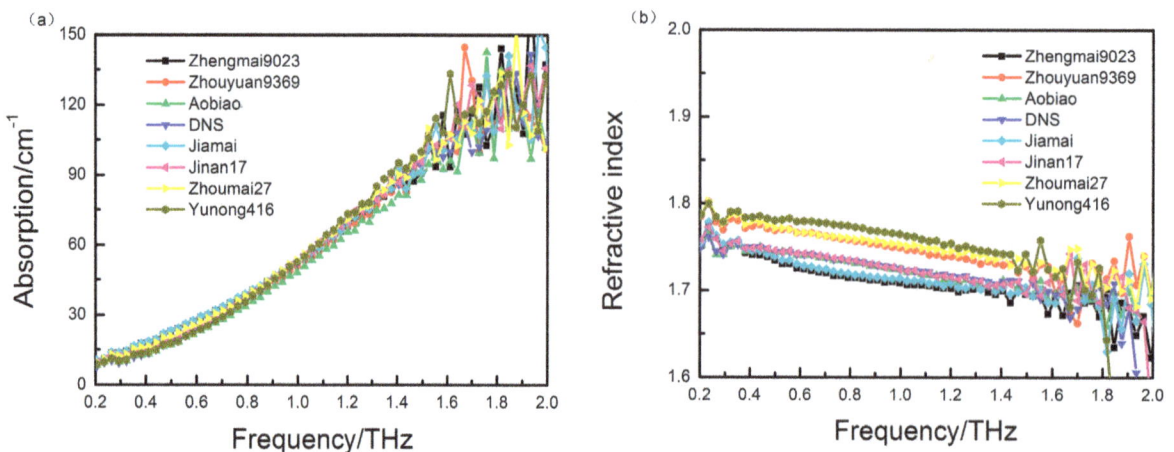

**Figure 2. (a)** The absorption coefficient and **(b)** refractive index for the eight wheat samples in the range of 0.2–2.0 THz.

As shown in Figure 1, the effective measurement range is 0.2–2.5 THz, and the sample spectra have narrow bandwidths. Although there is a minor difference in the shift among the samples, there is remarkable shift between the samples and the reference, which indicates that the refractive indices of the samples are different. Moreover, the amplitude changes of the samples indicate that the sample absorption coefficients differ.

Because of the similar components in the samples and the chemical complexity of wheat, the optical parameters of the measured wheat samples are very similar, as shown in Figure 2. In order to represent the variation in measurement effectively, the average absorption coefficients of 20 samples of eight wheat samples are shown in Figure 3. The average absorption coefficients are 37.1492 (zhengmai9023), 39.4354 (zhouyuan9369), 35.3358 (aobiao), 42.1133 (DNS), 37.5468 (jiamai), 42.2520 (jinan17), 39.8226 (zhoumai27) and 39.9409 (yunong416), respectively. In addition, the spectra of wheat samples at higher frequencies (above 1.5 THz) produce lower SNR due to the limitation of the dynamic range of the measurement system. Because there are no obvious absorption peaks in the spectra and because the differences in the absorption spectra for the eight wheat samples are not significant, we employ PLS regression to investigate the relationship between the minor spectral differences and the measured wheat varietal properties.

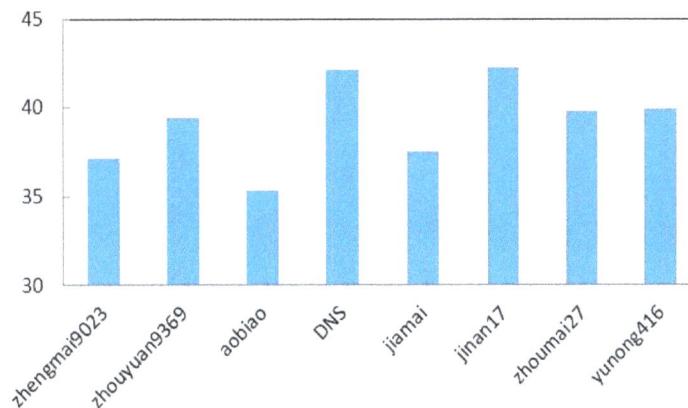

**Figure 3.** The average absorption coefficient of eight wheat samples.

## 3.2. PLS Analysis

The PLS analysis was performed on THz spectral data with the MATLAB software package (Version 2012a, Mathworks Inc., Natick, MA, USA) using the PLS toolbox (Version 4.0, Eigenvector Research Inc., Wenatchee, USA) and user-written scripts. The original spectra with 128 frequency variables were analyzed to establish the correlation between the spectral data and relevant wheat varieties. The eight wheat samples were set as output variables using numbers 1 through 8 (Table 1). The PLS calibration model was developed using a leave-one-out cross-validation calculation. In this study, both the absorption spectra and the refractive spectra of wheat samples from 0.2 to 1.5 THz were applied to obtain the best model. Then, the models were employed to predict the eight varieties of samples. A set of eight varieties of wheat (160 samples in total) was used in this experiment. All the samples were divided into two sets randomly, the calibration set (96 samples) and the prediction set (64 samples). The performance of the cross-validation models is shown in Table 2.

**Table 2.** Calibration and validation results obtained with the effective spectrum PLS Model.

| Input Variable | Frequency Range (THz) | Factors | Calibration | | Cross Validation | |
|---|---|---|---|---|---|---|
| | | | R | RMSEC | R | RMSECV |
| Absorption coefficient | 0.2–1.5 THz | 5 | 0.987 | 0.759 | 0.983 | 1.028 |
| Refractive index | 0.2–1.5 THz | 5 | 0.982 | 1.472 | 0.979 | 1.684 |

The absorption coefficient, which has a higher R-value and lower RMSE and RMSECV values, demonstrates better performance than the refractive index for the full-spectrum PLS model. Thus, this indicates that the absorption spectrum-based PLS model is a better model for prediction of wheat varieties compared to the refractive index-based PLS model in the frequency range of 0.2–1.5 THz.

Figure 4 shows the calibration and validation results for wheat varietal discrimination using the absorption spectrum PLS regression model. In the model, the reference line indicates the zero residuals between the predicted and the actual values. Figure 4 shows that the predicted values for all varieties of samples agree with the actual values, indicating that the PLS model can identify wheat varieties.

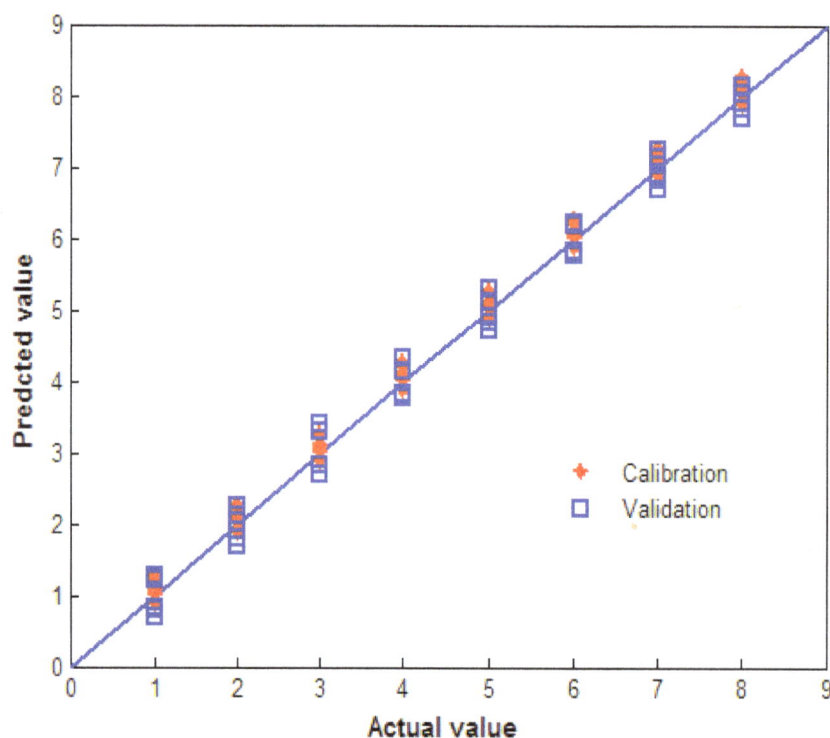

**Figure 4.** The calibration and validation results for wheat discrimination using the PLS model.

*3.3. iPLS Analysis*

For comparison, iPLS analysis was also performed using the same absorption spectral data sets to improve the model performance. First, the full spectrum was divided into 16 equal subintervals with eight variables. Calibration models were developed for each of the 16 intervals. Then, cross-validation was performed for each of the 16 models. Figure 5 presents the iPLS variable selection results for the discrimination of wheat varieties.

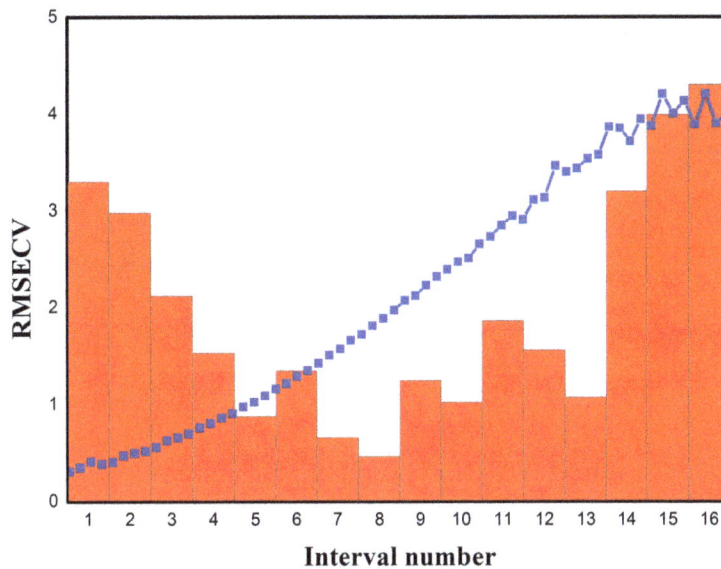

**Figure 5.** The iPLS results for the THz spectra data. The columns indicate the RMSECV in each subinterval, and the mean absorption spectrum of the wheat samples is overlaid on the plot.

From Figure 5, the width of each column is the same, while the height of each column indicates the RMSECV value calculated by the subinterval PLS model. The eighth column with the interval 57–64, corresponding to the frequency range of 787.5–900 GHz, shows the lowest RMSECV and is selected for developing the model. Therefore, the resulting model should have more precision in discriminating the wheat varieties.

In addition, iPLS selects the most relevant part of the spectrum, which can improve the performance of the model by removing the noise and interference from other regions. In the experiment, a different number of intervals were used to explore the best spectral region by the lowest RMSECV. The interval widths of 4 and 16 were also employed to construct the regression model, dividing the spectrum into 32 and 8 subintervals, respectively. Table 3 shows the calibration and validation results of the optimal iPLS models on THz absorption spectra.

**Table 3.** Calibration and validation results obtained with the optimal iPLS regression model.

| Interval Variables | Frequency Range | $R_{cal}$ | RMSEC | RMSECV |
|---|---|---|---|---|
| 4 | 0.731–0.956 THz | 0.991 | 0.768 | 1.260 |
| 8 | 0.787–0.900 THz | 0.992 | 0.573 | 0.967 |
| 16 | 0.675–1 THz | 0.984 | 0.837 | 1.237 |

As shown in Table 3, the iPLS model with eight interval variables has the lowest RMSECV value (0.967), the lowest RMSEC value (0.573), and the highest R value (0.992), compared to those of the full spectrum PLS model (Table 2). It is clear that the performance of the subinterval PLS model was better than the full-spectrum PLS model.

Figure 6 shows plots of the actual values compared to the values predicted by the PLS models based on the full THz absorption spectra and the optimal interval in the 0.787–0.9 THz range. The RMSEP of the full-spectrum PLS model is 0.845, while the RMSEP of the iPLS model is improved to 0.642.

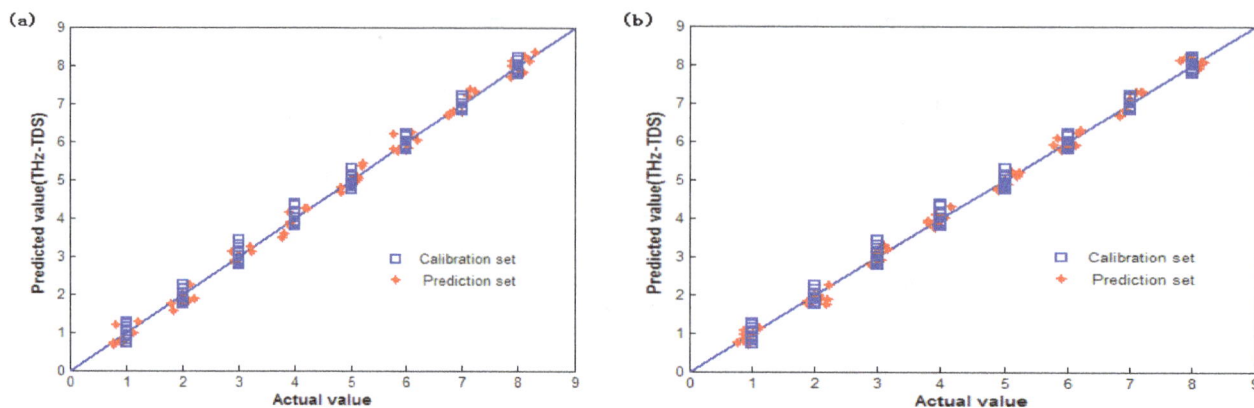

**Figure 6.** Scatter plots of the actual value *vs.* the predicted value for the discrimination of wheat varieties using (**a**) the full-spectrum PLS calibration model and (**b**) the iPLS calibration model based on the selected subinterval in the range 0.787–0.9 THz.

In Figure 6, the eight subintervals along the X-axis represent the eight wheat varieties as follows: intervals from 0.5–1.5, 1.5–2.5, 2.5–3.5, 3.5–4.5, 4.5–5.5, 5.5–6.5, 6.5–7.5, and 7.5–8.5 represent samples 1–8, respectively. The Y-axis represents the predicted model value for each wheat variety. The maximum range of values predicted by the PLS model for the discrimination wheat grains is 0.67–1.28, while that for the iPLS model is 0.70–1.15, which produces satisfactory results in the evaluation of the wheat varietals. The dispersion degree of the predicted value for Aobiao wheat and Zhengmai 9023 wheat is relatively high, as shown in Figure 6a, but both varieties can be discriminated correctly.

Comparing Figure 6a,b, improved prediction accuracy of the optimized PLS model for the discrimination of wheat varieties is observed, and the relative prediction error in the full-spectrum model is reduced. These results indicate that both the PLS and iPLS models can obtain good predictions for discriminating wheat varieties using THz spectroscopy; however, the iPLS model improves the prediction accuracy through optimal spectral region selection technology.

### 3.4. Discussion

Compared with other spectroscopic methods, such as NIR, MIR, and Raman, THz spectroscopy is a non-destructive rapid method to discriminate wheat varieties. The authors of [6] report that both MIR and NIR techniques can be applied in the industry as a rapid analytical tool to measure the quality parameters. In our experiment, the total time required for preparation, measurement, and analysis of wheat samples was within 5 min, which is also equal to that in [6,7] using data fusion with multiple analytical measurements to increase the model performance and decrease the modeling error for the prediction of quality parameters of crude oil, and the average RMSEP for the fully fused model (which included the IR, Raman, and NMR data) was calculated to be 0.307%. However, few studies compare the iPLS and PLS methods on the spectroscopic data of sample for the improvement in model performance.

Due to complex compositions in the wheat, the proportion of the characteristic components is small, and the overlapping signals obscure the THz absorption spectra. The absorption spectra for the wheat samples obtained in this experiment are featureless. In fact, featureless spectral absorption in the THz range that increases with increasing frequency is expected for many disordered amorphous materials [23–25]. Even though the characteristic absorption peaks are lacking, variations in the composition of different wheat samples generate changes in the THz absorption curves. The absorption spectra of various samples have markedly different THz spectra. Thus, chemometric analysis was employed using PLS regression to discriminate the eight wheat samples.

To obtain the best prediction model, both PLS and iPLS analysis methods were applied to construct regression models for discriminating the wheat varieties based on the THz absorption spectra. The above-mentioned analysis indicates that the model performance is primarily affected by the selection of the subinterval range. An appropriate interval width can provide the best prediction results. If the subinterval width is too large, noise and irrelevant information as well as the input sample spectra cause the model to predict the wheat varieties with lower accuracy.

Furthermore, correlations between neighboring regions in the spectrum cannot be obtained by the iPLS model [26]. In particular, the PLS model has only one subinterval (the full spectrum), which includes all spectral information without removing the noise or irrelevant intervals. Thus, the PLS performance is not better than the iPLS model performance. If the interval width is too low, and the corresponding subinterval is too broad, high prediction error will be caused by the loss of small signature information in the spectrum.

Moreover, the prediction accuracy of the constructed model is also affected by the THz absorption of sample. The measured THz spectra of sample are dependent on the experimental environment, experimental setup, background noise, uncertainty of the time-domain signal system [27], and sample preparation process. Moreover, the chemical constituents of wheat grains are determined by many factors [28], such as the wheat variety, plant environment, climate, harvest year, and so on. Even though wheat samples from different harvest years are prepared, the absorption spectra of samples are likely different from each other due to the difference between the chemical compositions of samples. To construct an accurate PLS regression model for wheat varietal discrimination using THz absorption spectra, the influencing factors of the spectra, such as wheat samples from different harvest years in the same plant area and wheat samples from different harvest years and different plant areas, will need to be considered and explored in further systemic studies.

In our work, to avoid the influence of these factors on the performance of the prediction model, the average spectra of the samples were produced by repeated measurements under the same conditions. The PLS regression model was developed to discriminate the wheat varieties based on the THz spectra, and iPLS was used to remove irrelevant information from the spectra. The RMSECV and RMSEC values were used to evaluate the performance of the PLS model. The best subinterval in the iPLS model, which was determined using the lowest RMSECV, was selected for the PLS calibration model. The prediction results for the wheat varieties were improved by using the iPLS model with eight interval variables and the lowest RMSECV value of 0.967.

## 4. Conclusions

In this study, the potential of THz spectra as a new, non-destructive technology for wheat varietal discrimination has been demonstrated. First, the THz spectra from 0.2 to 2 THz were measured and analyzed for eight wheat varieties using the transmission configuration. Then, PLS regression was employed to obtain a prediction model for wheat varieties. For comparison, iPLS was also employed to improve the performance of the calibration model using the proper subinterval in the spectrum. The results showed that good prediction models could be obtained with lower RMSECV and higher R-values in the two regression methods. In particular, the prediction accuracy for wheat grain varieties was improved with the iPLS calibration model ($R = 0.992$ and RMSECV $= 0.967$). As a result, THz spectroscopy associated with chemometric techniques showed useful measurement and discrimination of wheat varieties. However, further studies should be considered to account for the large number of mixed varieties and for varieties harvested in different years.

## Acknowledgments

This work was supported by the National High-Tech R&D Program of China (863 Program) (Grant No. 2012AA101608) and the National Natural Science Foundation of China (Grant No. 61071197).

## Author Contributions

In this paper, Hongyi Ge, Yuying Jiang and Feiyu Lian conducted the algorithm design, experiments and analysis. Technical discussion and data processing were performed by Hongyi Ge and Yuying Jiang. Yuan Zhang and Shanhong Xia have supervised the whole work. All authors contributed to the writing and the revision of the manuscript.

## Conflicts of Interest

The authors declare no conflict of interest.

## References

1. Zapotoczny, P., Discrimination of wheat grain varieties using image analysis: Morphological features. *Eur. Food Res. Technol.* **2011**, *233*, 769–779.

2. Pourreza, A.; Pourreza, H.; Abbaspour-Fard, M.H.; Sadrnia, H., Identification of nine Iranian wheat seed varieties by textural analysis with image processing. *Comput. Electron. Agric.* **2012**, *83*, 102–108.

3. Levandi, T.; Pussa, T.; Vaher, M.; Ingver, A.; Koppel, R.; Kaljurand, M. Principal component analysis of HPLC-MS/MS patterns of wheat (*Triticum aestivum*) varieties. *Proc. Est. Acad. Sci.* **2014**, *63*, 86–92.

4. Paliwal, J.; Visen, N.S.; Jayas, D.S.; White, N.D.G. Cereal grain and dockage identification using machine vision. *Biosyst. Eng.* **2003**, *85*, 51–57.

5. Miralbes, C. Discrimination of European wheat varieties using near infrared reflectance spectroscopy. *Food Chem.* **2008**, *106*, 386–389.

6. Ferreira, D.S.; Galao, O.F.; Pallone, J.A.L.; Poppi, R.J. Comparison and application of near-infrared (NIR) and mid-infrared (MIR) spectroscopy for determination of quality parameters in soybean samples. *Food Control* **2014**, *35*, 227–232.

7. Dearing, T.I.; Thompson, W.J.; Rechsteiner, C.E.; Marquardt, B.J. Characterization of Crude Oil Products Using Data Fusion of Process Raman, Infrared, and Nuclear Magnetic Resonance (NMR) Spectra. *Appl. Spectrosc.* **2011**, *65*, 181–186.

8. Ferguson, B; Zhang, X.C. Materials for terahertz science and technology. *Nat. Mater.* **2002**, *1*, 26–33.

9. Vieira, F.S.; Pasquini, C. Determination of Cellulose Crystallinity by Terahertz-Time Domain Spectroscopy. *Anal. Chem.* **2014**, *86*, 3780–3786.

10. Melinger, J.S.; Laman, N.; Grischkowsky, D. The underlying terahertz vibrational spectrum of explosives solids. *Appl. Phys. Lett.* **2008**, *93*, doi:10.1063/1.2949068.

11. Liang, M.Y.; Shen, J.L.; Wang, G.Q. Identification of illicit drugs by using SOM neural networks. *J. Phys. D Appl. Phys.* **2008**, *41*, doi:10.1088/0022-3727/41/13/135306.

*12.* Ueno, Y.; Rungsawang, R.; Tomita, I.; Ajito, K. Quantitative measurements of amino acids by terahertz time-domain transmission spectroscopy. *Anal. Chem.***2006**, *78*, 5424–5428.

13. Ge, H.Y.; Jiang, Y.Y.; Xu, Z.H.; Lian, F.Y.; Zhang, Y.; Xia, S.H. Identification of wheat quality using THz spectrum. *Opt. Express* **2014**, *22*, 12533–12544.

14. Pupeza, I.; Wilk, R.; Koch, M. Highly accurate optical material parameter determination with THz time-domain spectroscopy. *Opt. Express* **2007**, *15*, 4335–4350.

15. Xiao-li, Z.; Jiu-sheng, L. Diagnostic techniques of talc powder in flour based on the THz spectroscopy. *J. Phys. Conf. Ser.* **2011**, *276*, doi:10.1088/0022-3727/41/13/135306.

16. Zhang, Y.; Peng, X.H.; Chen, Y.; Chen, J.; Curioni, A.; Andreoni, W.; Nayak, S.K.; Zhang, X.C. A first principle study of terahertz (THz) spectra of acephate. *Chem. Phys. Lett.* **2008**, *452*, 59–66.

17. Dorney, T.D.; Baraniuk, R.G.; Mittleman, D.M. Material parameter estimation with terahertz time-domain spectroscopy. *J. Opt. Soc. Am. A* **2001**, *18*, 1562–1571.

18. Duvillaret, L.; Garet, F.; Coutaz, J.L. A reliable method for extraction of material parameters in terahertz time-domain spectroscopy. *IEEE J. Sel. Top. Quantum Electron.* **1996**, *2*, 739–746.

19. Nicolai, B.M.; Beullens, K.; Bobelyn, E.; Peirs, A.; Saeys, W.; Theron, K.I.; Lammertyn, J. Nondestructive measurement of fruit and vegetable quality by means of NIR spectroscopy: A review. *Postharvest Biol. Technol.* **2007**, *46*, 99–118.

20. Norgaard, L.; Saudland, A.; Wagner, J.; Nielsen, J.P.; Munck, L.; Engelsen, S.B. Interval partial least-squares regression (iPLS): A comparative chemometric study with an example from near-infrared spectroscopy. *Appl. Spectrosc.* **2000**, *54*, 413–419.

21. Kachrimanis, K.; Braun, D.E.; Griesser, U.J. Quantitative analysis of paracetamol polymorphs in powder mixtures by FT-Raman spectroscopy and PLS regression. *J. Pharm. Biomed.* **2007**, *43*, 407–412.

22. Hua, Y.F.; Zhang, H.J. Qualitative and Quantitative Detection of Pesticides with Terahertz Time-Domain Spectroscopy. *IEEE Trans. Microw. Theory Tech.* **2010**, *58*, 2064–2070.

23. Naftaly, M.; Miles, R.E. Terahertz time-domain spectroscopy for material characterization. *IEEE Proc.* **2007**, *95*, 1658–1665.

24. Zhang, F.; Kambara, O.; Tominaga, K.; Nishizawa, J.; Sasaki, T.; Wang, H.W.; Hayashi, M. Analysis of vibrational spectra of solid-state adenine and adenosine in the terahertz region. *R. Soc. Chem. Adv.* **2014**, *4*, 269–278.

25. McIntosh, A.I.; Yang, B.; Goldup, S.M.; Watkinson, M.; Donnan, R.S. Crystallization of amorphous lactose at high humidity studied by terahertz time domain spectroscopy. *Chem. Phys. Lett.* **2013**, *558*, 104–108.

26. Andersen, C.M.; Bro, R. Variable selection in regression-a tutorial. *J. Chemometr.* **2010**, *24*, 728–737.

27. Withayachumnankul, W.; Fischer, B.M.; Lin, H.Y.; Abbott, D. Uncertainty in terahertz time-domain spectroscopy measurement. *J. Opt. Soc. Am. B* **2008**, *25*, 1059–1072.

28. Triboi, E.; Abad, A.; Michelena, A.; Lloveras, J.; Ollier, J.L.; Daniel, C. Environmental effects on the quality of two wheat genotypes: 1. quantitative and qualitative variation of storage proteins. *Eur. J. Agron.* **2000**, *13*, 47–64.

# 10

# Real-Time Human Pose Estimation and Gesture Recognition from Depth Images Using Superpixels and SVM Classifier

**Hanguen Kim** [1], **Sangwon Lee** [1], **Dongsung Lee** [2], **Soonmin Choi** [2], **Jinsun Ju** [2] **and Hyun Myung** [1,*]

[1] Urban Robotics Laboratory (URL), Dept. Civil and Environmental Engineering, Korea Advanced Institute of Science and Technology (KAIST), 291 Daehak-ro, Yuseong-gu, Daejeon 305-338, Korea; E-Mails: sskhk05@kaist.ac.kr (H.K.); lsw618@gmail.com (S.L.)

[2] Image & Video Research Group, Samsung S1 Cooperation, 168 S1 Building, Soonhwa-dong, Joong-gu, Seoul 100-773, Korea; E-Mails: dslee.lee@samsung.com (D.L.); soonmin.choi@samsung.com (S.C.); jinsun.ju@samsung.com (J.J.)

* Author to whom correspondence should be addressed; E-Mail: hmyung@kaist.ac.kr

Academic Editor: Assefa M. Melesse

**Abstract:** In this paper, we present human pose estimation and gesture recognition algorithms that use only depth information. The proposed methods are designed to be operated with only a CPU (central processing unit), so that the algorithm can be operated on a low-cost platform, such as an embedded board. The human pose estimation method is based on an SVM (support vector machine) and superpixels without prior knowledge of a human body model. In the gesture recognition method, gestures are recognized from the pose information of a human body. To recognize gestures regardless of motion speed, the proposed method utilizes the keyframe extraction method. Gesture recognition is performed by comparing input keyframes with keyframes in registered gestures. The gesture yielding the smallest comparison error is chosen as a recognized gesture. To prevent recognition of gestures when a person performs a gesture that is not registered, we derive the maximum allowable comparison errors by comparing each registered gesture with the other gestures. We evaluated our method using a dataset that we generated. The experiment results show that our method performs fairly well and is applicable in real environments.

**Keywords:** human pose estimation; gesture recognition; depth information; low-cost platform

## 1. Introduction

Human pose estimation and gesture recognition are attractive research topics in computer vision and robotics owing to their many applications, including human computer interaction, game control and surveillance. The release of low-cost depth sensors, such as Microsoft Kinect for Xbox 360 and ASUS Xtion, has provided many important benefits to these research areas [1]. Kinect for Xbox 360 and Xtion are RGB-D (red, green, blue and depth) sensors that obtain depth information by structured light technology [2]. The structured light sensors infer the depth values by projecting an infrared light pattern onto a scene and analyzing the distortion of the projected light pattern. However, these sensors are limited to indoor use, and their low resolution and noisy depth information make it difficult to estimate human poses from depth images. Many human pose estimation methods use a GPU (graphic processing unit) to increase the frame rate and the performance [3–6]. These methods shows remarkable performance, but it is difficult to operate the algorithms on low-cost systems, such as embedded boards or mobile platforms. Other methods that do not use GPUs show low frame rates [7,8], and some cannot even run in real time [9]. Moreover, model-based approaches require model calibration before pose estimation [7,8].

Human pose estimation methods can be classified into two categories: model-based and learning-based approaches. In model-based approaches, prior knowledge of a human body model is required, and the human pose is estimated by inverting the kinematics or solving optimization problems. Grest *et al.* exploit the iterative closest point (ICP) approach with a body model to track a human pose initialized by a hashing method [8]. Siddiqui *et al.* use the Markov chain Monte Carlo (MCMC) framework with head, hand and forearm detectors to fit a body model [10]. Zhang *et al.* introduced a generative sampling algorithm with a refinement step of local optimization with a 3D body model for body pose tracking [3]. In learning-based approaches, however, a human body model is not considered, but human poses are directly estimated from input images with various machine learning algorithms. Shotton *et al.* present two different methods for human pose estimation [4]. The methods are based on a random forest trained on a large amount of synthetic human depth image data. One of the methods uses a per-pixel classification method, where each pixel on a human body is classified by the trained classification random forest. The other method predicts joint position by using a regression random forest. Each pixel on the human body directly votes on all of the joint positions. The classified pixels or the joint position votes are aggregated to estimate joint points by a mean shift. Hernández-Vela *et al.* extended the per-pixel classification method of Shotton *et al.* [4] using graph-cut optimization [5]. The graph-cut is an energy minimization framework, and it has been widely used in image segmentations.

Various methods are currently used for gesture recognition, including RGB-D sensor-based methods and other sensor-based (e.g., inertial measurement unit, electromyography, virtual reality gloves, *etc.*) methods [11]. Since the launch of Kinect for Xbox 360, many studies on gesture recognition use the skeleton information provided by Kinect for Xbox 360 [12,13] or directly use depth information [14,15]. However, with most of the algorithms, registration of arbitrary gestures that users may perform is not easy, because most of the algorithm use machine learning-based approaches that require a training process [13,16]. Furthermore, the recognition rate is easily affected by environmental changes [17].

Gesture recognition methods can be divided into two categories: matching-based and machine learning-based approaches. Wu *et al.* proposed a matching-based method that uses dynamic time-warping to identify users and recognize gestures with joint data from Kinect for Xbox 360 [12]. Megavannan *et al.* also proposed a matching-based algorithm that uses the motion dynamics of an object from the depth difference and average depth information [14]. The performance of matching-based approaches is easily affected by external noise or environmental changes, but training data and training phase are not required. The machine learning-based methods require training data and a training phase to generate classifiers, but they are more robust to noise and environmental changes than matching-based methods. Biswas *et al.* presented a method wherein SVMs are trained to classify gestures with depth difference information [13]. Sigalas *et al.* proposed an upper-body part tracking method and a gesture recognition method that combines a multi-layer perceptron and radial basis function neural networks [18].

In this paper, we propose human pose estimation and gesture recognition algorithms that use only depth information for robustness to environmental and lighting changes. The proposed algorithms are designed to be operated on low-cost systems, such as embedded boards and mobile platforms, without exploiting GPUs. Our pose estimation method is based on a per-pixel classification method where each pixel on the human body is classified into a body part. We reduce the computation time of body part classification by using superpixels. The proposed human pose estimation method can estimate human poses instantly without a calibration process, allowing the system to be used with any subject immediately. In the proposed gesture recognition method, the gesture registration process is simple, and gestures can be recognized regardless of motion speed by using key frame extraction. The proposed gesture recognition method is robust to environmental or lighting changes, as it uses only pose information, and our method can cover various motions for a single gesture by adapting the Mahalanobis distance [19] in comparing input motions with the registered gesture.

The remainder of this paper is organized as follows: Section 2 explains our human pose estimation method. Section 3 describes the proposed gesture recognition method. The proposed algorithms are evaluated through experiments in Section 4. Finally, a conclusion and directions for future work are provided in Section 5.

## 2. Human Pose Estimation

The proposed human pose estimation method is based on an SVM (support vector machine) and superpixels. Our pose estimation method predicts 15 joint positions of the human body: head, neck, torso, L/R (left/right) shoulders, L/R elbows, L/R hands, L/R hips, L/R knees and L/R feet. A flow diagram of the proposed pose estimation algorithm is presented in Figure 1. The proposed system extracts background-subtracted human depth ROIs (regions of interest), ensuring that each extracted ROI contains a human body without occlusion. Once a human body ROI is extracted from a depth image, the depth values of the pixels on the human body are normalized. The normalization process starts from finding the minimum and maximum depth values in the extracted ROI image. Then, the depth values are linearly mapped into 16-bit scale by using the minimum and maximum depth values. By normalization of depth values in the extracted ROI image, the ROI has the same depth value distribution regardless of the distance from the depth sensor. Furthermore, the normalization process stretches the range of the depth values in the ROI, so that small pixel differences in depth images become more distinguishable. After normalization, superpixels are generated on the human body using SLIC (simple linear iterative clustering) [20] for speeding up computation time. After superpixels are generated, the origin of $x, y$ coordinates of superpixels to the central moment of the human body is to keep the coordinate points consistent regardless of the distance between the depth sensor and the human body, which is beneficial to the performance of SVM. The converted superpixels are then scaled to a predefined body size depending on the depth values of the human body. The processed superpixels are classified by a trained SVM classifier into one of the body parts, and the falsely-classified superpixels are removed in the optimization process. From the classified superpixels, the positions of the 15 joints are estimated. When hands are situated on or over the torso, they are not estimated from the classifier. In this case, the hands are tracked in the torso area using a Kalman filter. The overall procedures are presented in Figure 1.

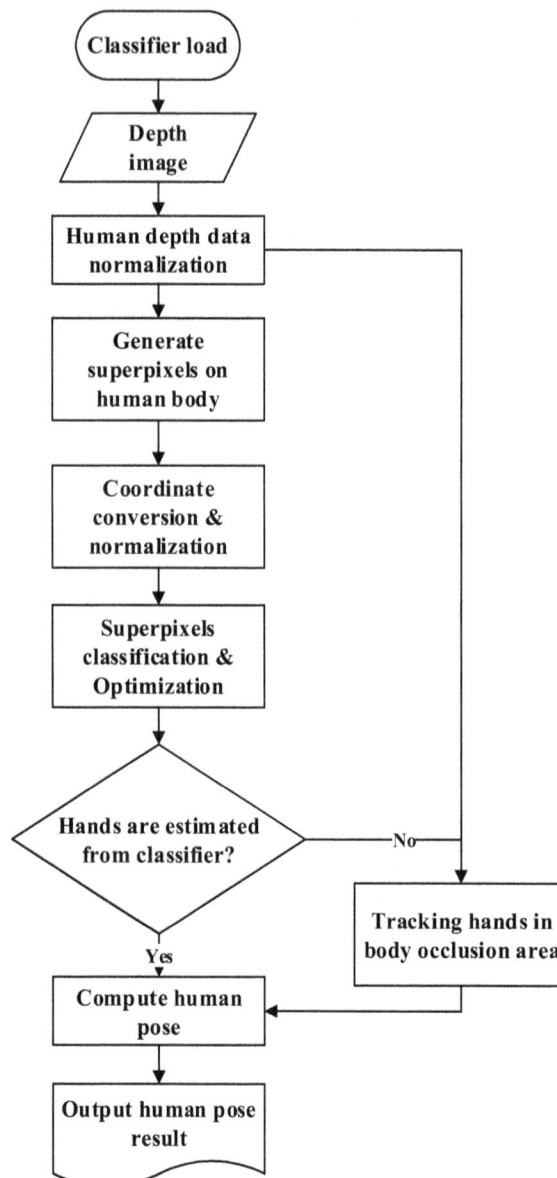

**Figure 1.** Flow diagram of the proposed human pose estimation method.

## 2.1. Superpixel Feature Generation

Our pose estimation method uses per-pixel classification without prior human model information. However, a great deal of time is required to classify all of the pixels in a human ROI. To address this problem, instead of classifying all of the pixels in the human ROI, superpixels are generated on the human body, and then, the generated superpixels are classified into one of the body parts. This process tremendously reduced the amount of time required for body part classification. For the generation of superpixels, the SLIC (simple linear iterative clustering) superpixels method [20] is exploited. According to [20], the superpixels algorithm groups pixels into perceptually-meaningful atomic regions. The SLIC superpixels method adopts k-mean clustering to generate superpixels. The original superpixel algorithm iteratively clusters the pixels with similar color intensity within the user-defined search space. The proposed algorithm, however, generates a constant number of superpixels by using the depth values, which is to obtain the pose estimation within constant computation time regardless of the distance between the depth sensor and the human body. In this paper, the number of superpixels was empirically adjusted. The proposed method helps to keep the computation time for the human pose estimation almost constant. The exemplary results of superpixel generation on the human body are shown in Figure 2.

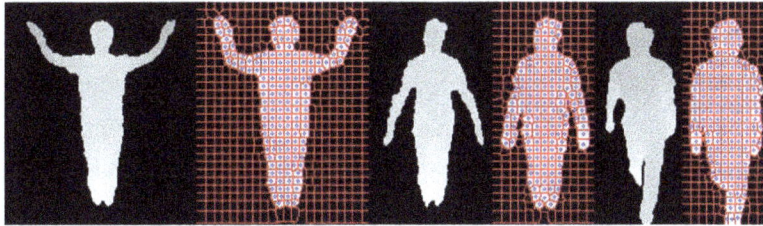

**Figure 2.** Results of superpixel generation on human bodies. The blue points on the human bodies are the generated superpixels.

As the origin of the $x, y$ coordinates in the human body ROI is the upper left vertex of the ROI and the width and the height of the ROI change with respect to the human pose, the $x, y$ coordinates of the same body part vary with the human pose. To reduce these variations of the $x, y$ coordinates, the origin of the $x, y$ coordinates of superpixels is translated to the central moment of the human body. In addition to the coordinate conversion, the variation of the $x, y$ coordinates of the superpixels caused by the distance change between the human body and the depth sensor should be considered. The variation of the $x, y$ coordinates from the distance change is mitigated by multiplying the $x, y$ coordinates by the scale factor $S$ computed from the following equation:

$$S = \frac{D_{body}}{D_{ref}} \tag{1}$$

where $D_{body}$ and $D_{ref}$ denote the average depth of the human body and the reference depth value, respectively. Equation (1) is derived on the basis of the relationship between the distance and the height of objects in an image being inverse-linear. Equation (1) means that the human body is placed at the reference distance $D_{ref}$. Based on the specification of Kinect for Xbox 360 or ASUS Xtion Pro [21], the $D_{ref}$ can be set between 2000 and 4000 mm, where depth sensors give less noisy depth values and human body size is appropriate in the images.

### 2.1.1. Pose Estimation

The joint positions are estimated from the classified superpixels. Our human pose estimation method uses the SVM for classification of superpixel features. The SVM classifier learns from the training data that have been created by a motion capture system, and the performance of the classifier is verified by the ground truth data that have been also produced by the motion capture system. The motion capture system used in this paper will be explained in the Experiments Section. If there are some misclassified superpixels, the estimated joint positions may result in wrong positions. To prevent this, a misclassified superpixel is removed in the optimization process. When every superpixel is correctly classified, the superpixels that belong to the same body part are clustered on the corresponding body part. From this, we can assume that the misclassified superpixels are located far from the cluster of the corresponding body part. Therefore, by measuring the distance between the same body part superpixels, the misclassified superpixels can be identified and removed. We define $B = \{B_1, B_2, ..., B_i, ..., B_N\}$, a vector for body part labels whose components $B_i$ indicate a certain body part label, $P = \{P_1, P_2, ..., P_j, ..., P_M\}$, a set of the generated superpixels, and $b = \{b_1, b_2, ..., b_j, ..., b_M\}$, a vector for labels of the superpixels whose components $b_j$ specify the body part label assigned to a superpixel $P_j$. Algorithm 1 shows the detailed procedure for removing the mislabeled superpixels. For a superpixel $P_j$ that belongs to a body part $B_i$, $meanDist_{B_i,P_j}$, the mean distance to the other superpixels, classified as $B_i$, is computed. For the body part $B_i$, $meanDist_{B_i}$, the mean of $meanDist_{B_i,P_j}$, is computed. After computing the $meanDist_{B_i,P_j}$ and $meanDist_{B_i}$, every $meanDist_{B_i,P_j}$ is compared with $meanDist_{B_i}$. If $meanDist_{B_i,P_j}$ is bigger than $meanDist_{B_i}$, the superpixel $P_j$ will be removed, because the superpixel $P_j$ is situated away from the cluster of superpixels classified as $B_i$. The overall procedures can be found in Algorithm 1. Figure 3a,b present examples of misclassified superpixels and the optimized label results, respectively.

---

**Algorithm 1** Removal of mislabeled superpixels

1: **for** each body part label $B_i \in B$ **do**
2:    **for** each superpixel $P_j(b_j = B_i)$ **do**
3:       $meanDist_{B_i,P_j} \leftarrow$ compute the mean distance to $P_k((b_k = B_i)\&(j \neq k))$
4:    **end for**
5:    $meanDist_{B_i} \leftarrow$ compute the mean of $meanDist_{B_i,P}$
6: **end for**
7: **for** each body part label $B_i \in B$ **do**
8:    **for** each superpixel $P_j(b_j = B_i)$ **do**
9:       **if** $meanDist_{B_i,P_j} > meanDist_{B_i}$ **then**
10:          $b_j \leftarrow$ (none)
11:       **end if**
12:    **end for**
13: **end for**

---

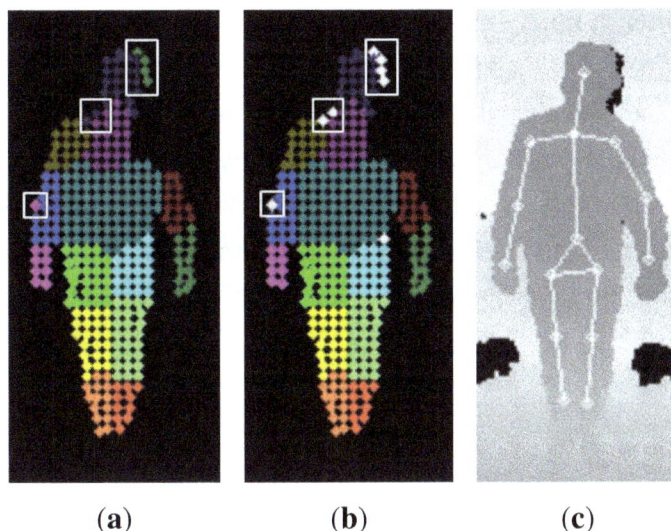

**Figure 3.** An example of: (**a**) superpixel classification; (**b**) optimization; (**c**) pose estimation. Misclassified superpixels removed in the optimization process are indicated by white rectangles.

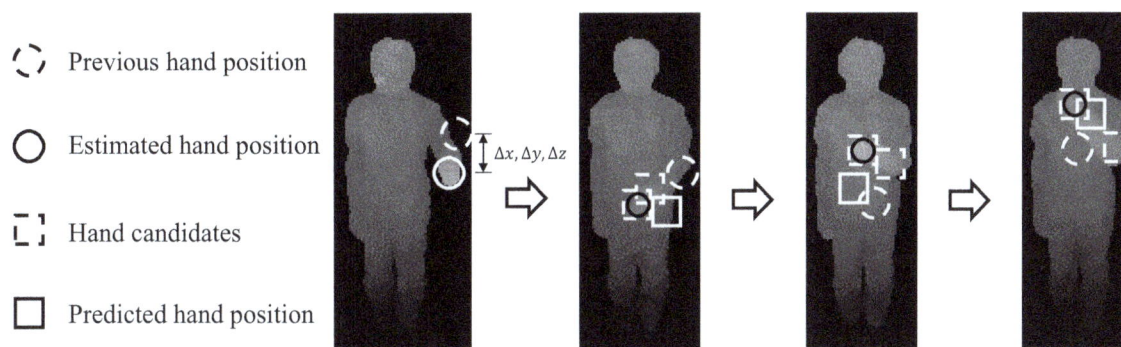

**Figure 4.** Example of the measurement update step when the hand occludes the torso. The hand tracker extracts the depth measurements for hand candidates. The final hand position is estimated by the depth measurement with the smallest Mahalanobis distance from the previous hand position.

After removing the misclassified superpixels, each joint position of $B_i$ is estimated as the central moment of the superpixels labeled as the corresponding body part $B_i$. An example of a pose estimation result is shown in Figure 3c. However, when the hands occlude the torso, none of the superpixels are classified as hands. To solve this problem, a hand tracker is designed to estimate the hand position, even when the hand information is not provided by the classifier. The hand tracker is designed based on the Kalman filter. The Kalman filter usually consists of two steps. One is the state prediction step, and the other is the measurement update step, which are calculated at every frame in the background process. In the state prediction step, the state is estimated based on the previous hand position and the hand position difference, *i.e.*, $\Delta x, \Delta y$ and $\Delta z$. In the measurement update step, the hand tracker extracts the depth measurements for hand candidates within the ROI (region of interest), which is calculated from the previous hand position and $\Delta x, \Delta y$ and $\Delta z$. The hand position is updated by the depth measurement with the smallest Mahalanobis distance from the previous hand position. If the hand position can be acquired from the classification results, the hand position is corrected by the classification results. Otherwise, the result from the Kalman filter is finally used as the hand position. The exemplary measurement update procedure is shown in Figure 4 in case the hand occludes the torso. The rationale for applying the linear Kalman filter as a hand tracker is as follows. The first reason is that the hand movements are continuous. Therefore, the hand position can be predicted by using the previous hand position and its position difference. The second reason is that the hand movement being faster than the operation speed of the overall algorithm (in our experimental setting, the overall algorithm runs at 15 frames per second) was not considered. This means that the hand tracker can track the general hand movements continuously within the operation speed.

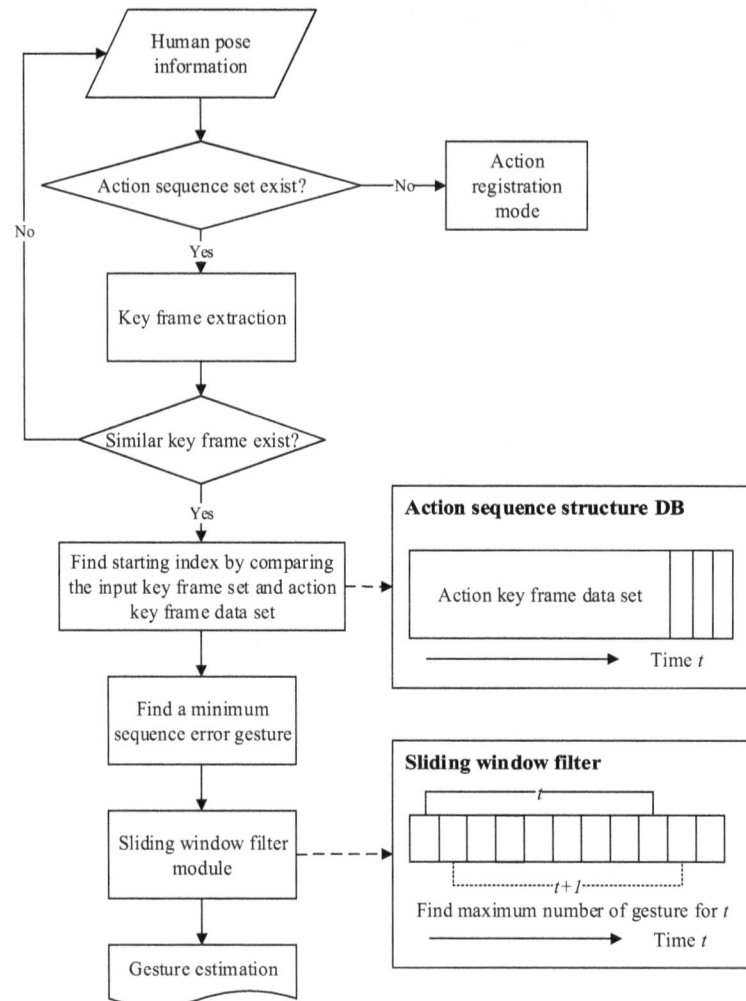

**Figure 5.** Flow diagram of the proposed gesture recognition method.

## 3. Gesture Recognition

The proposed gesture recognition method recognizes gestures by inspecting the pose information from our pose estimation algorithm. The flow diagram of the proposed gesture recognition method is shown in Figure 5. When the pose information is given at each frame, key frames are extracted from the given sequence of pose information. The input pose information is normalized to the specific size for a robust recognition regardless of the body type. The key frames are important frames of the gestures, and they are extracted when the pose information difference between the previous key frame and the current frame is above the pre-defined threshold. Several key frames are extracted until the currently-extracted key frame is similar to one of the previously-generated key frames. The extracted key frames are then compared with key frames of registered gestures. The gesture with the smallest comparison error is chosen as the estimated gesture. The estimated gesture is then passed through the window filter to prevent misrecognition of gestures.

### 3.1. Key Frame Extraction

The human pose estimation result cannot be directly used for gesture recognition, because the joint positions vary depending on the body shape and the distance to the depth sensor. To reduce such effects, the joint information should be normalized to a predefined body size. By applying the normalization process, the joint information fits into a predefined body size.

The key frame extraction is a fundamental part of the proposed gesture recognition method. For each frame, the estimated joint positions are compared with the previous key frame. If the joint position differences between the previous key frame and the current frame are above the pre-defined threshold, the current frame is stored as a key frame. With only consideration of the pose information difference in the key frame extraction process, the same key frames can be extracted for the same gesture regardless of the motion speed. The key frame extraction continues until the currently extracted the key frame is similar to one of the previously-extracted key frames. When the key frame extraction is completed, normalized joint positions of the extracted key frames are stored as an "action sequence." Figure 6 shows an example of an action sequence. An action sequence contains a sequence of key pose frames for a certain human body motion.

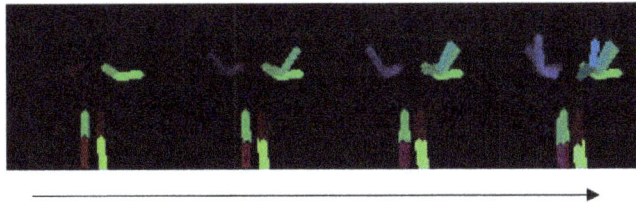

**Figure 6.** An example of an action sequence.

### 3.2. Action Sequence Matching

The gestures are recognized by comparing an input action sequence with the action sequences of registered gestures. A registered gesture contains its name and an action sequence that describes the motions of the gesture. The action sequence in the registered gesture has key frames, and in each key frame, means and standard deviations of joint positions are stored. The registered gesture that gives the smallest comparison error is chosen as the estimated gesture. In the comparison of the input action sequence with registered gesture action sequences, however, the first key frame of the input action sequence does not always match that in the action sequences of the registered gestures, because the key frames are extracted without information about the start and end time of a gesture. We define the start key frame $k_{Sg}$ in the action sequence of the registered gesture $g$ as the keyframe that matches the first key frame of the input action sequence, and it is found by the following equation:

$$k_{Sg} = \operatorname*{arg\,min}_{k\in\{0,...,K-1\}} \frac{1}{N} \sum_{n=0}^{N-1} \frac{(f_{0,n} - m_{g,k,n})^2}{\sigma_{g,k,n}^2} \tag{2}$$

where $f_{0,n}$ is the $n$-th joint position in the first key frame of the input action sequence, $m_{g,k,n}$ is the mean of the $n$-th joint position in the $k$-th key frame of the registered gesture action sequence $g$, $\sigma_{g,k,n}$ is the standard deviation of the $n$-th joint position in the $k$-th key frame of the registered gesture action sequence $g$, $K$ is the total number of key frames in the registered gesture action sequence and $N$ is the number of human body joints. Equation (2) delineates that the start key frame in the registered action sequence has the smallest comparison error, which is defined as the mean Mahalanobis distance of joint positions in the first key frame in the input action sequence.

Before the recognition of gestures, gestures that we want to recognize should be defined. In the gesture registration process, a gesture to be registered should be acted several times. In each gesture action, the poses of the gesture should be slightly different so that various movements in a single gesture can be covered. From the generated several action sequences for a single gesture, means and variances of joint positions in each key frame of the gesture are computed. These means and variances of joint positions are used to compute a comparison error with an input action sequence. In addition to the means and variances of the action sequences, maximum comparison errors should be considered to prevent undefined gestures from being recognized as one of the registered gestures and to prevent wrong estimation of gestures. A maximum comparison error of a gesture is set as the smallest comparison error between the other registered gestures. The maximum comparison error of a registered gesture $g$ is computed as follows:

$$E_{\max} = \min_{\substack{g'\in\{0,...,G-1(g'\neq g)\}, \\ a\in\{0,...,A-1\}}} \frac{1}{K} \sum_{k=0}^{K-1} \frac{1}{N} \sum_{n=0}^{N-1} \frac{(f_{a,k,n} - m_{g',k',n})^2}{\sigma_{g',k',n}^2}, k' = (k_{Sg'} + k) \bmod K \tag{3}$$

where $f_{a,k,n}$ is the $n$-th joint position in the $k$-th key frame of the $a$-th action sequence from the registered gesture $g$; $m_{g',k',n}$ is the mean of the $n$-th joint position in the $k'$-th key frame of the other registered gestures $g'$; $\sigma_{g',k',n}$ is the standard deviation of the $n$-th joint position in the $k'$-th key frame of the other registered gesture $g'$; $G$ is the number of registered gestures; and $A$ is the number of action sequences in the registered gesture $g$. The meaning of Equation (3) delineates that the maximum comparison error of a gesture is the smallest comparison error from every case of comparison between the several action sequences of gesture $g$ and the other gestures $g'$. After determining the start key frame in the registered gesture action sequence $g$, the input action sequence and the registered gesture action sequence $g$ are compared in a circular manner, as shown in Figure 7. For each registered gesture action sequence, the start key frame is found, and the comparison error with the input action sequence is computed using the following equation:

$$E_{\text{compare}} = \frac{1}{K_I} \sum_{k=0}^{K_I-1} \frac{1}{N} \sum_{n=0}^{N-1} \frac{(f_{k,n} - m_{g,k',n})^2}{\sigma_{g,k',n}^2}, k' = (k_{Sg} + k) \bmod K \tag{4}$$

where $K_I$ is the number of key frames in the input action sequence and $f_{k,n}$ is the $n$-th joint position in the $k$-th key frame of the input action sequence. Equation (4) indicates that the comparison error of a registered gesture with the input action sequence is the mean Mahalanobis distance of joint positions in the key frames from the input action sequence.

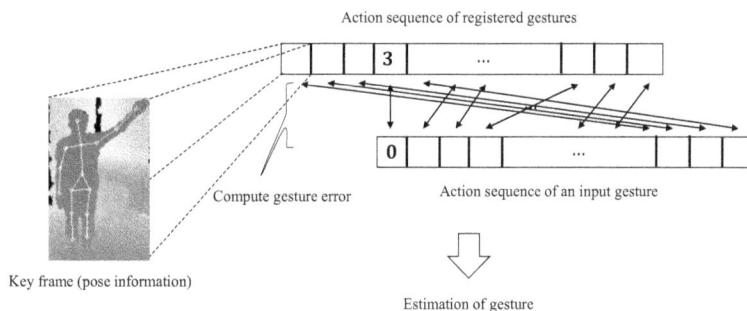

**Figure 7.** Action sequence matching.

When computation of the comparison error for every registered gesture is done, the gesture with the smallest comparison error is selected as a candidate for a recognized gesture. If the comparison error is below the maximum comparison error of the gesture, the selected gesture undergoes the sliding window filter to minimize misrecognition of gestures. Every time when the input action sequence is estimated, the sliding window filter with a pre-defined length is applied to the sequence. If the count of the same gesture estimation exceeds the pre-defined ratio of the window size, the gesture is accepted as the final gesture recognition result.

## 4. Experiments

In this section, the proposed human pose estimation and gesture recognition method are evaluated. The evaluation is performed on datasets that we created. The quantitative results of the human pose estimation method are given, and the proposed method is compared with OpenNI [22] and Shotton's algorithm [4]. The gesture recognition method is evaluated by defining five gestures and testing the proposed method with 50 datasets for each gesture. The experiments are performed on a PC with an Intel i5 3.0 GHz quad-core CPU with 4-GB RAM. The average computation times of the proposed methods are shown in Table 1. A depth sensor ASUS Xtion pro [21] is used to create the test datasets.

**Table 1.** Average computation times of the proposed methods.

|      | Pose estimation | Gesture recognition | Total frame rate |
|------|-----------------|---------------------|------------------|
| Time | 65 ms           | less than 1 ms      | 15 fps           |

The test datasets are captured with data from a depth sensor Xtion pro installed at about a 2-m height, and the size of the captured images is $320 \times 240$. The datasets consist of 30 video files containing about 11kframes in total. The ground truth data that contain the joint positions at each frame are created for every test dataset using a commercial motion capture system, iPi Mocap Studio [23]. The iPi Mocap Studio is a scalable markerless motion capture system that uses 3D depth sensors to track human joints and produce 3D animations. This system provides human pose data with centimeter-level accuracy offline.

The performance of the proposed human pose estimation method is evaluated by computing the error metrics with ground truth data. The error metrics of the pose estimation algorithm in OpenNI

NiTE [22] are also computed and compared with the proposed method. Table 2 shows the error metrics of the proposed method and OpenNI NiTE. The results of OpenNI NiTE show better performance in the accuracy of the pose estimation, but the performance difference of the two methods is small enough. On the other hand, the pose estimation of OpenNI NiTE has the disadvantage that it requires a calibration process that should be performed in the initial stage. Table 3 shows the average initial human pose estimation time of the proposed method and OpenNI NiTE. The initial human pose estimation time is determined by taking the time difference between the first pose estimation and the first input frame with the test dataset. In OpenNI NiTE, the calibration process requires users to take a pre-defined calibration pose for a moment, which sometimes takes more than five seconds. Only after the calibration process, OpenNI NiTE can estimate the poses of users. However, the proposed method does not require a calibration process, and human poses are immediately determined from depth images. These results show that the proposed human pose estimation algorithm is more suitable for a gesture recognition algorithm that requires fast recognition performance. We further tried to compare the human pose estimation efficiency with Shotton's algorithm [4], as shown in Table 4. The comparison was performed by using the performance result of Shotton's paper [4]. The table shows the computational cost for human pose estimation per each frame. According to the comparison result, the computation time of the proposed algorithm requires only about 56% that of Shotton's algorithm. Therefore, the proposed method is more suitable for an implementation in the embedded surveillance system. Figure 8 shows examples of experimental results obtained with the proposed pose estimation method.

**Table 2.** Experimental results of pose estimation. L, left; R, right.

| Body part | Mean (mm) | | SD (mm) | |
|---|---|---|---|---|
| | Proposed method | OpenNINiTE | Proposed method | OpenNI NiTE |
| Head | 26.0 | 26.0 | 25.2 | 15.5 |
| Neck | 53.3 | 37.5 | 28.4 | 26.4 |
| Torso | 75.3 | 121.0 | 31.7 | 32.8 |
| L Shoulder | 41.2 | 45.7 | 21.9 | 24.6 |
| L Elbow | 86.2 | 80.1 | 69.2 | 66.4 |
| L Hand | 199.4 | 128.2 | 332.6 | 156.1 |
| R Shoulder | 49.6 | 34.3 | 20.9 | 23.2 |
| R Elbow | 87.7 | 68.9 | 61.9 | 52.2 |
| R Hand | 190.8 | 97.4 | 306.7 | 120.2 |
| L Hip | 44.2 | 34.9 | 29.0 | 27.2 |
| L Knee | 133.0 | 44.3 | 44.3 | 24.6 |
| L Foot | 114.0 | 68.7 | 28.6 | 41.3 |
| R Hip | 58.2 | 43.9 | 23.8 | 29.6 |
| R Knee | 130.0 | 52.6 | 47.0 | 26.2 |
| R Foot | 96.8 | 81.6 | 28.0 | 74.4 |
| Average | 92.3 | 64.3 | 73.3 | 48.8 |

**Table 3.** Average initial human pose estimation time (unit: ms).

| | Proposed Method | OpenNI NiTE |
|---|---|---|
| Time | 67.0 | 2413.1 |

**Table 4.** Computational cost of human pose estimation (unit: GFlops).

| | Proposed method | Shotton's Algorithm [4] |
|---|---|---|
| Computational cost for each frame | 0.81 | 1.44 |

(**a**)

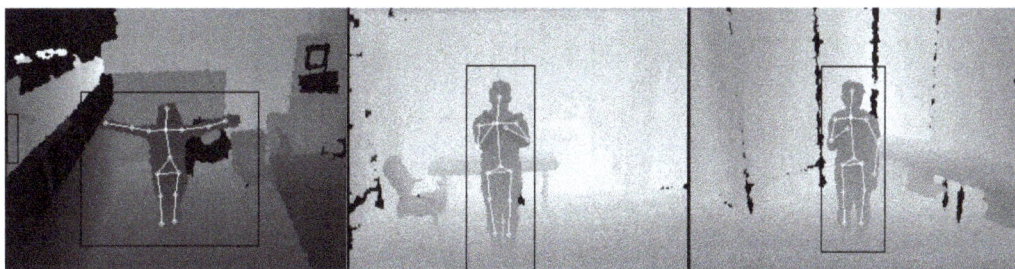

(**b**)

**Figure 8.** Experimental results of pose estimation. (**a**) Superpixels classification results; (**b**) pose estimation results.

In the case of the proposed gesture recognition algorithm, we defined five gestures to evaluate the proposed gesture recognition method: "request for help", "emergency", "request for emergency supplies", "complete" and "suspension of work". "Request for help" is defined as a motion where a person is beating his/her chest with a single hand. "Emergency" is an action of waving a single hand in the air. "Request for emergency supplies" is a motion that resembles someone lifting something with a single hand. The "complete" gesture is making a circle with raised arms. Lastly, "suspension of work" is a motion of crossing arms. We assumed that all gestures were made toward the sensor within the human pose estimation range ($\pm30$ degrees). This assumption may not guarantee a good gesture recognition rate depending on the viewing angle of the sensor. This problem, however, can be mitigated by installing multiple sensors for a wider viewing angle. For each gesture, 50 test datasets that recorded the behavior of people with various body types (thin, normal and overweight ratio 1:3:1), genders (male and female ratio 4:1) and clothes (casual, protective clothing and suit clothing) were used to evaluate the performance of the proposed gesture recognition method. If a gesture is recognized correctly once or more than once in a test video, it is counted as a correct recognition; a true positive case. If no gesture is recognized or gestures are recognized incorrectly, it is counted as a false recognition; a false positive case. Table 5 shows the experimental results of the proposed algorithm, and Table 6 shows the false recognition cases of the gestures. The experimental results show that the false recognition rate of 'emergency supplies' is higher than that of any other gesture. This result may be attributed to the similarity of the motion of the 'emergency supplies' gesture to the partial motion of the other gestures. The experimental results of the proposed gesture recognition method can be improved if we define gestures where motions of the gesture have minimal overlap with each other. Figure 9 shows example experimental results of the proposed gesture recognition method. As a result, using the test datasets that recorded various people, the human pose-based gesture recognition algorithm can recognize the gesture robustly.

**Table 5.** Experimental results of gesture recognition.

| Gesture | Recognition | False Recognition | Recognition Rate | False Recognition Rate |
|---|---|---|---|---|
| "Request for help" | 47 | 11 | 94.0% | 5.5% |
| "Emergency" | 43 | 4 | 86.0% | 2.0% |
| "Request for emergency supplies" | 49 | 24 | 98.0% | 12.0% |
| "Complete" | 46 | 7 | 92.0% | 3.5% |
| "Suspension of work" | 47 | 1 | 94.0% | 0.5% |
| Total | 232 | 47 | 92.8 | 4.7 |

**Table 6.** False gesture recognition results.

| Gesture Dataset | False Recognition Cases | | | | |
|---|---|---|---|---|---|
| | "Request for Help" | "Emergency" | "Request for Emergency Supplies" | "Complete" | "Suspension of Work" |
| "Request for help" | None | 0 | 0 | 1 | 0 |
| "Emergency" | 0 | None | 5 | 0 | 0 |
| "Request for emergency supplies" | 1 | 4 | None | 0 | 0 |
| "Complete" | 0 | 0 | 5 | None | 1 |
| "Suspension of work" | 10 | 0 | 14 | 6 | None |
| Total | 11 | 4 | 24 | 7 | 1 |

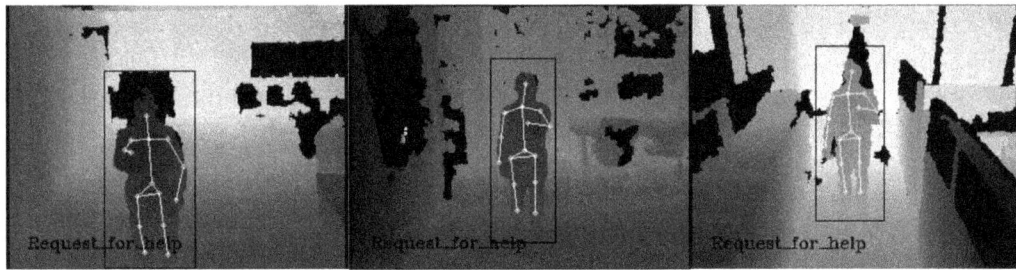

(a)

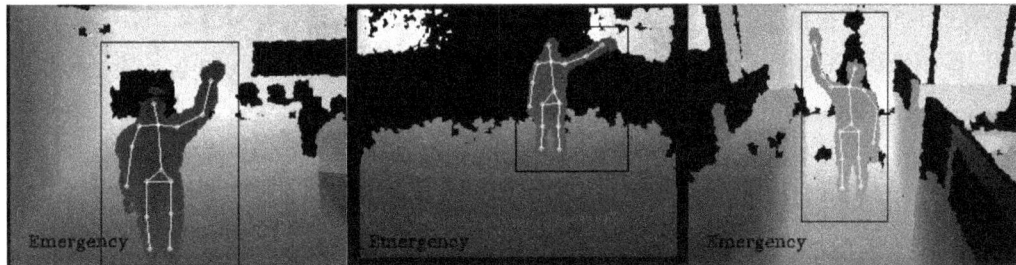

(b)

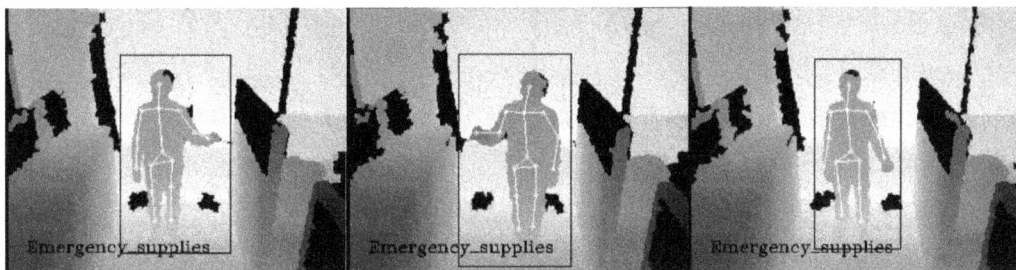

(c)

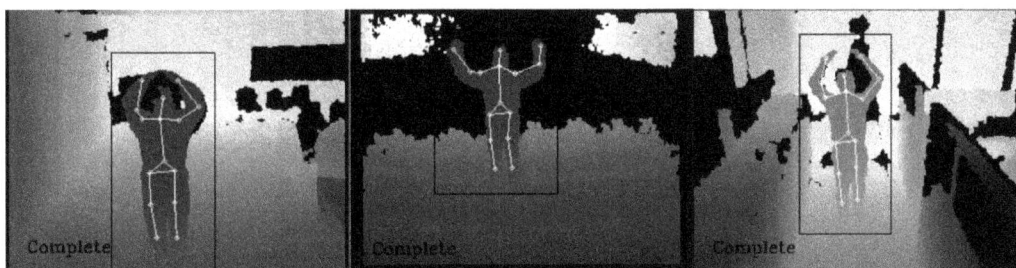

(d)

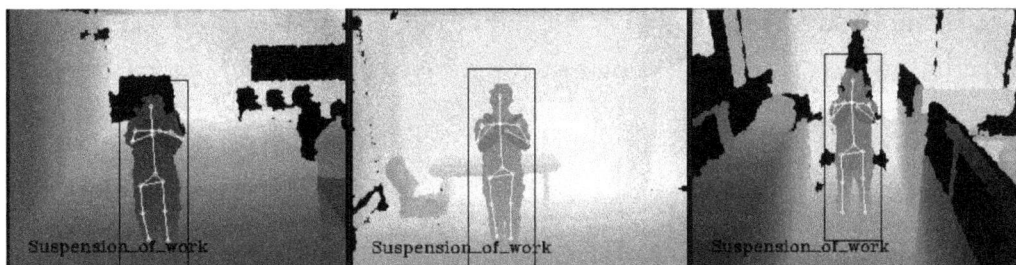

(e)

**Figure 9.** Example experimental results of gesture recognition. (**a**) The experimental results: "request for help"; (**b**) the experimental results: "emergency"; (**c**) the experimental results: "request for emergency supplies"; (**d**) the experimental results: "complete"; (**e**) the experimental results: "suspension of work".

## 5. Conclusions

In this work, we proposed a human pose estimation and a gesture recognition method with a depth sensor. In the human pose estimation, joint positions of a human body are estimated only with depth information, and the proposed method can be operated on low-cost platforms without exploiting GPUs. Our pose estimation method is based on per-pixel classification and superpixels. Instead of classifying all of the pixels on a human body, superpixels are generated on a human body, and then, the generated superpixels are classified into one of the body parts. This process greatly reduces the computation time. In gesture recognition, the pose information from the pose estimation method is used to extract key frames. A set of extracted key frames is compared with registered gestures, and a predicted result is passed to the sliding window filter. The key frame extraction enables robust and fast gesture recognition regardless of motion speed. The experimental results show that the proposed human pose estimation and gesture recognition method provide acceptable performance in real environment applications.

The current methods use SVMs for the classification of superpixels, and training data are generated manually. The performance of the current method can be improved by dealing with the following topics.

- Employing other classification algorithms for body part classification, such as a deep learning method.
- Applying kinematic constraints of a human body in pose estimation.
- Estimating human orientation to compensate for human orientation change in the human pose estimation process.
- Using better feature information for body part classification.

## Acknowledgments

This research was financially supported by Samsung S1 Cooperation. The students were supported by the Ministry of Land, Infrastructure and Transport (MoLIT) as the U-City Master and Doctor Course Grant Program.

## Author Contributions

Hanguen Kim and Sangwon Lee conducted the algorithm design, experiments and analysis under the supervision of Hyun Myung. Dongsung Lee, Soonmin Choi, and Jinsun Ju produced test datasets and conducted experiments and analysis. The authors were involved in writing the paper, the literature review and the discussion of the results.

## Conflicts of Interest

The authors declare no conflict of interest.

# References

1. Zhang, Z. Microsoft Kinect sensor and its effect. *IEEE Multimed.* **2012**, *19*, 4–10.
2. Arieli, Y.; Freedman, B.; Machline, M.; Shpunt, A. Depth Mapping Using Projected Patterns. U.S. Patent 8,150,142, 3 April 2012.
3. Zhang, Z.; Soon Seah, H.; Kwang Quah, C.; Sun, J. GPU-accelerated real-time tracking of full-body motion with multi-layer search. *IEEE Trans. Multimed.* **2013**, *15*, 106–119.
4. Shotton, J.; Girshick, R.; Fitzgibbon, A.; Sharp, T.; Cook, M.; Finocchio, M.; Moore, R.; Kohli, P.; Criminisi, A.; Kipman, A.; *et al.* Efficient human pose estimation from single depth images. *IEEE Trans. Pattern Anal. Mach. Intell. (PAMI)* **2013**, *35*, 2821–2840.
5. Hernández-Vela, A.; Zlateva, N.; Marinov, A.; Reyes, M.; Radeva, P.; Dimov, D.; Escalera, S. Graph Cuts Optimization for Multi-limb Human Segmentation in Depth Maps. In Proceedings of IEEE Conference on Computer Vision and Pattern Recognition (CVPR), St. Paul, MN, USA, 14–18 May 2012; pp. 726–732.
6. Ganapathi, V.; Plagemann, C.; Koller, D.; Thrun, S. Real Time Motion Capture Using a Single Time-of-flight Camera. In Proceedings of IEEE Conference on Computer Vision and Pattern Recognition (CVPR), San Francisco, CA, USA, 13–18 June 2010; pp. 755–762.
7. Zhu, Y.; Fujimura, K. Constrained Optimization for Human Pose Estimation from Depth Sequences. In Proceedings of Asian Conference on Computer Vision (ACCV), Tokyo, Japan, 1–22 November 2007; pp. 408–418.
8. Grest, D.; Woetzel, J.; Koch, R. Nonlinear Body Pose Estimation from Depth Images. In *Pattern Recognition*; Springer: Berlin/Heidelberg, Germany, 2005; pp. 285–292.
9. Ye, M.; Wang, X.; Yang, R.; Ren, L.; Pollefeys, M. Accurate 3D Pose Estimation from a Single Depth Image. In Proceedings of IEEE International Conference on Computer Vision (ICCV), Barcelona, Spain, 6–13 November 2011; pp. 731–738.
10. Siddiqui, M.; Medioni, G. Human Pose Estimation from a Single View Point, Real-time Range Sensor. In Proceedings of IEEE Computer Society Conference on Computer Vision and Pattern Recognition Workshops (CVPRW), San Francisco, CA, USA, 13–18 June 2010; pp. 1–8.
11. Bellucci, A.; Malizia, A.; Diaz, P.; Aedo, I. Human-display interaction technology: Emerging remote interfaces for pervasive display environments. *IEEE Pervasive Comput.* **2010**, *9*, 72–76.
12. Wu, J.; Konrad, J.; Ishwar, P. Dynamic Time Warping for Gesture-based User Identification and Authentication with Kinect. In Proceedings of IEEE International Conference on Acoustics, Speech and Signal Processing (ICASSP), Vancouver, BC, Canada, 26–31 May 2013; pp. 2371–2375.
13. Biswas, K.; Basu, S.K. Gesture Recognition Using Microsoft Kinect®. In Proceedings of International Conference on Automation, Robotics and Applications (ICARA), Wellington, New Zealand, 6–8 December 2011; pp. 100–103.
14. Megavannan, V.; Agarwal, B.; Babu, R.V. Human Action Recognition Using Depth Maps. In Proceedings of International Conference on Signal Processing and Communications (SPCOM), Bangalore, India, 22–25 July 2012; pp. 1–5.

15. Kim, H.; Hong, S.H.; Myung, H. Gesture recognition algorithm for moving Kinect sensor. In Proceedings of IEEE International Workshop on Robots and Human Interactive Communications (RO-MAN), Gyeongju, Korea, 26–29 August 2013; pp. 320–321.

16. Xia, L.; Chen, C.C.; Aggarwal, J. View Invariant Human Action Recognition Using Histograms of 3D Joints. In Proceedings of IEEE Computer Society Conference on Computer Vision and Pattern Recognition Workshops (CVPRW), Providence, RI, USA, 16–21 June 2012; pp. 20–27.

17. Poppe, R. A survey on vision-based human action recognition. *Image Vision Comput.* **2010**, *28*, 976–990.

18. Sigalas, M.; Baltzakis, H.; Trahanias, P. Gesture Recognition based on Arm Tracking for Human-robot Interaction. In Proceedings of IEEE/RSJ International Conference on Intelligent Robots and Systems (IROS), Taipei, Taiwan, 18–22 October 2010; pp. 5424–5429.

19. De Maesschalck, R.; Jouan-Rimbaud, D.; Massart, D.L. The mahalanobis distance. *Chemom. Intell. Lab. Syst.* **2000**, *50*, 1–18.

20. Achanta, R.; Shaji, A.; Smith, K.; Lucchi, A.; Fua, P.; Susstrunk, S. SLIC superpixels compared to state-of-the-art superpixel methods. *IEEE Trans. Pattern Anal. Mach. Intell. (PAMI)* **2012**, *34*, 2274–2282.

21. ASUS Xtion pro. Available online: http://www.asus.com/Multimedia/Xtion_PRO/ (accessed on 26 February 2015).

22. OpenNI NiTE. Available online: http://www.openni.ru/files/nite/ (accessed on 26 February 2015).

23. IPi Mocap Studio. Available online: http://ipisoft.com/ (accessed on 26 February 2015).

# 11

# Assessing and Correcting Topographic Effects on Forest Canopy Height Retrieval Using Airborne LiDAR Data

**Zhugeng Duan** [1,2,3]**, Dan Zhao** [1]**, Yuan Zeng** [1,*]**, Yujin Zhao** [1]**, Bingfang Wu** [1] **and Jianjun Zhu** [2]

[1] Key Laboratory of Digital Earth Science, Institute of Remote Sensing and Digital Earth (RADI), Chinese Academy of Science, Haidian District, Beijing 100094, China; E-Mails: dzg47336628@163.com (Z.D.); zhaodan@radi.ac.cn (D.Z.); zhaoyj@radi.ac.cn (Y.Z.); wubf@radi.ac.cn (B.W.)

[2] School of GeoSciences and Info-Physics, Central South University, Changsha 410083, China; E-Mail: zjj@mail.csu.edu.cn

[3] School of Sciences, Central South University of Forestry and Technology, Changsha 410004, China

* Author to whom correspondence should be addressed; E-Mail: zengyuan@radi.ac.cn

Academic Editor: Assefa M. Melesse

**Abstract:** Topography affects forest canopy height retrieval based on airborne Light Detection and Ranging (LiDAR) data a lot. This paper proposes a method for correcting deviations caused by topography based on individual tree crown segmentation. The point cloud of an individual tree was extracted according to crown boundaries of isolated individual trees from digital orthophoto maps (DOMs). Normalized canopy height was calculated by subtracting the elevation of centres of gravity from the elevation of point cloud. First, individual tree crown boundaries are obtained by carrying out segmentation on the DOM. Second, point clouds of the individual trees are extracted based on the boundaries. Third, precise DEM is derived from the point cloud which is classified by a multi-scale curvature classification algorithm. Finally, a height weighted correction method is applied to correct the topological effects. The method is applied to LiDAR data acquired in South China, and its effectiveness is tested using 41 field survey plots. The results show that the terrain impacts the canopy height of individual trees in that the downslope side of the tree trunk is elevated and the upslope side is depressed. This further affects the extraction of the location and crown of individual trees. A strong correlation was detected between the slope gradient and the proportions of returns with height differences more than

0.3, 0.5 and 0.8 m in the total returns, with coefficient of determination $R^2$ of 0.83, 0.76, and 0.60 (n = 41), respectively.

**Keywords:** LiDAR; canopy height point cloud; Dinghushan National Nature Reserve; multi-resolution segmentation; topography

---

## 1. Introduction

Airborne light detection and ranging (LiDAR) has become an effective and reliable way to map terrain and retrieve forest structural parameters [1]. In fact, forest managers have found LiDAR to be of great utility when compared with traditional methods as a way to obtain forest information. Many important forest parameters can be obtained directly or indirectly from LiDAR data, such as tree height, crown width, diameter, canopy density, volume and biomass [2–4]. Before forest parameters obtained by LiDAR inversion can be applied, understanding the causes and magnitude of errors of such parameters is essential. A variety of factors can cause errors in LiDAR-based estimates including the terrain, forest structure and point cloud filtering algorithms; variations in topography also play a key role in data extraction.

A series of studies have shown the accuracy of LiDAR-derived digital elevation models (DEMs) and tree parameters generally decreases as slope gradient increases. When the slope gradient increases from 15.6° to 37.6°, the vertical Root Mean Square Error of tree height extraction increases from 0.576 m to 0.901 m [5]. Hodgson and Bresnahan [6] examined the effects of topography on tree height and spatial structure of forest within a small plot. Gatziolis *et al.* [7] studied the accuracy of an airborne LiDAR-derived DEM in a coniferous forest area with high biomass. Their results showed that DEM accuracy was mainly affected by the ground slope and sensor accuracy; increasing slope gradient resulted in reduced tree height because LiDAR inversion and DEM errors led to forest volume errors.

Breidenbach *et al.* [8] investigated tree height retrieved from LiDAR and InSAR data and concluded that slopes generally impacted the estimation of tree height. They also suggested that as the slope gradient increased, models neglecting slopes would overestimate tree height, and noted that this could be corrected by a slope coefficient that would allow a more accurate estimation of tree height.

Complex forest habitats and systems require accurate DEM extraction as the basis for forest parameter inversion. The accuracy of DEM extraction in turn is related to ground point classification and DEM interpolation. Complex terrains tend to cause misclassifications and missed points; therefore, the ground point cloud does not always reflect the actual terrain conditions, resulting in lower DEM accuracy, thus affecting parameter extraction [9]. Evans *et al.* [4] studied the effect of ground point misclassifications on the extraction of vegetation height. Interpolation is typically required to generate a DEM. However, the precision of the interpolation is again dependent on terrain complexity [10]. Much attention has been placed on the impact of DEM interpolation on the extraction of canopy height and tree height. Filtering algorithms used to generate LiDAR-based DEMs for complex forests mainly include Iterative Approximation [11–13], Progressive Densification [14], Morphological Filtering [15–17], and Multi-scale Curvature Classification (MCC) [18].

Forest canopy height describes the top of the vegetated canopy as well as the vertical and horizontal distribution of the canopy; it is the key to assessing forest parameters [2,19]. An accurate estimate of canopy height significantly affects the inversion of forest parameters. Canopy height can be used to directly measure individual tree parameters such as crown vertices, crown diameter, and height to the first live branch [20]. Additionally, canopy height can also be used to calculate the Diameter at Breast Height (DBH) of individual trees, the average DBH of a forest stand, as well as stand volume and biomass through relative growth equations. In a word, canopy height is the basis for individual tree parameters such as tree height, crown diameter, height to the first live branch, DBH, canopy density, and other parameters including parameters related to forest structure, volume and biomass. Therefore, the accurate extraction of canopy height is the key to accurate estimation of forest structure parameters and biomass retrieval. Two methods used to express canopy height are the discrete normalized point cloud method and grid canopy height models (CHM). The former is obtained directly by subtracting DEM data from filtered and classified point cloud data, while the latter was produced by subtracting DEM data from a Digital Surface Model (DSM); a DSM can be generated by interpolating filtered and classified point cloud data [21,22]. Therefore, a CHM is also known as a Normalized Digital Surface Model (nDSM), Digital Canopy Height Model (DCHM) or Digital Canopy Model (DCM) [23].

Both the normalized point cloud and CHMs are subject to the influence of terrain. Previous researchers have paid a considerable amount of attention to the influence of terrain on DEMs and the impact of classifications and filtering algorithms on forest parameter extraction, but few studies have explored the process of obtaining normalized point cloud data or CHMs. Canopy height differences are generated by subtracting a DEM from a DSM or the raw point cloud data. This paper proposes a method for terrain correction of canopy height based on individual tree crown segmentation. The aim is to assess and correct topographical effects on forest canopy height retrieval using airborne LiDAR data.

## 2. Study Area and Data

### 2.1. Study Area

The study area (Figure 1) is located in Dinghushan National Nature Reserve (23°05', 23°15'N, 112°30', 112°57'E), Dinghu District of Zhaoqing City, in the west central part of Guangdong Province, China. The area has a mean elevation of 545 m (minimum, 14.1 m; maximum, 1000.3 m). A rather complex terrain characterises the topography of the study area. The mountains in this hilly area run downhill from southwest to northeast. The area is covered by steep slopes, among which 44.7% are between 0–5°; 2.0%, 5–8°; 3.3%, 8–5°; 6.8%, 15–25°; 17.1%, 25–35°; and 26.1% are ≥35°. Vegetation in the area falls into six categories: evergreen broadleaf forest, coniferous and broadleaf mixed forest (here after referred to as "mixed forest"), tropical evergreen coniferous forest, montane evergreen shrub, montane evergreen bushes, and anthropogenic vegetation. The main tree species in the study area were *Pinus massoniana*, *Schima superba*, *Castanopsis chinensis*, and *Eucalyptus robusta*.

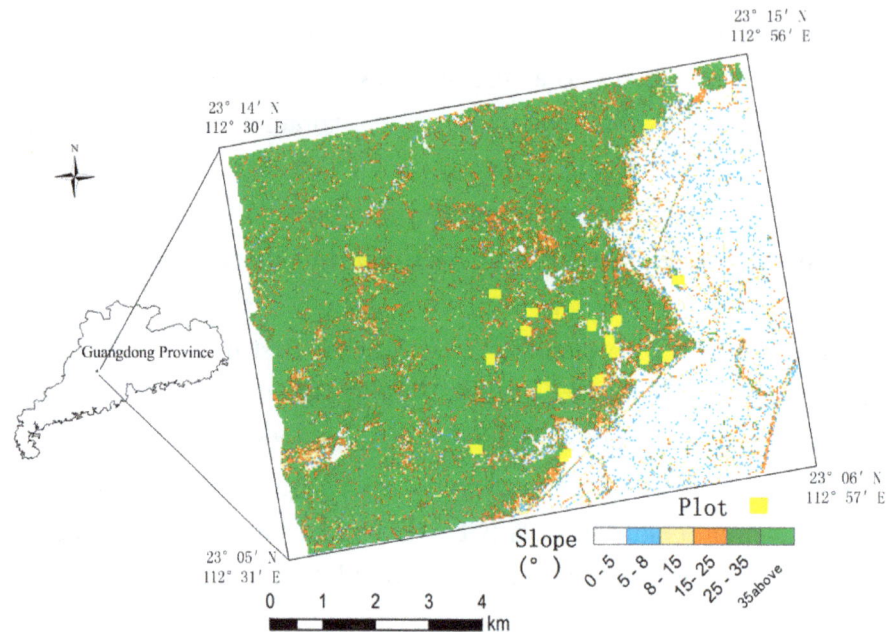

**Figure 1.** Distribution of slope in study area and plots.

## 2.2. LiDAR Data

The LiDAR survey was carried out on 24 December 2012 using LiteMapper 6800 (Ingenieur Geshellshaft für Interfaces mbH (IGI), Kreuztal, Germany, www.igi.eu); the main technical parameters are presented in Table 1. LiteMapper 6800 is capable of providing full waveform LiDAR data and high-resolution aerial images simultaneously because it is equipped with Rollei PRO 6.5 megapixels digital cameras (Rollei, Braunschweig, Germany). A Y-5 aircraft was used as an aerial platform. The aircraft covered a total area of 60 km$^2$, with an average speed of 160 km/h, an absolute cruising altitude of 1300 m, an average relative altitude of 750 m, and an average point cloud density of 15.0 p/m$^2$.

**Table 1.** Main technical parameters of LiteMapper 6800.

| Device Type | LiteMapper 6800 |
|---|---|
| Pulse repetition frequency | Up to 400 KHz |
| Laser wavelength | 1550 ns |
| Pulse length | 3.5 ns |
| Laser beam divergence | ≤0.5 mrad |
| Multiple target separation within single shot | 0.6 m |
| Return pulse width resolution | 0.15 m |
| Scan pattern | Parallel scan |
| Scan angle range | ±30° |
| Angle readout resolution | 0.001° |
| Ground sample spot diameter | 0.24 m (@800 m) |
| Horizontal accuracy | 0.08 m (@800 m) |
| Vertical accuracy | 0.04 m (@800 m) |

## 2.3. Field Inventory Data

The ground-truth data were collected from 16 November 2012 to 9 December 2012. A total of 41 plots (30 × 30 m$^2$), with most slopes between 8–40° (Figure 2a), were established, including four in coniferous forests, 22 in broadleaf forests, and 15 in mixed forests. Parameters measured mainly included: DBH of individual trees with a DBH >5 cm (DBH is measured from 1.3 m above ground), tree height, height to the first live branch, crown diameter, canopy density, slope gradient and aspect; all served as samples and evidence for LiDAR-based biomass inversion and biodiversity research. Plot area was measured by a forest compass combined with a measuring tape. Slope gradients were also determined by a forest compass. Angular point coordinates of plots were determined by the wide area differential signals of a Trimble3000 handheld GPS (Trimble, Sunnyvale, CA, USA), with sub-meter nominal accuracy. Individual tree DBHs were measured by a DBH tape; tree height and height to the first live branch were measured by a laser altimeter; crown semidiameter was measured in east-west and north-south directions with a measuring tape. Figure 2b shows that the average crown semidiameter in the plots was 1.6–5.2 m, and maximum crown semidiameter was 3.3–25.0 m, respectively.

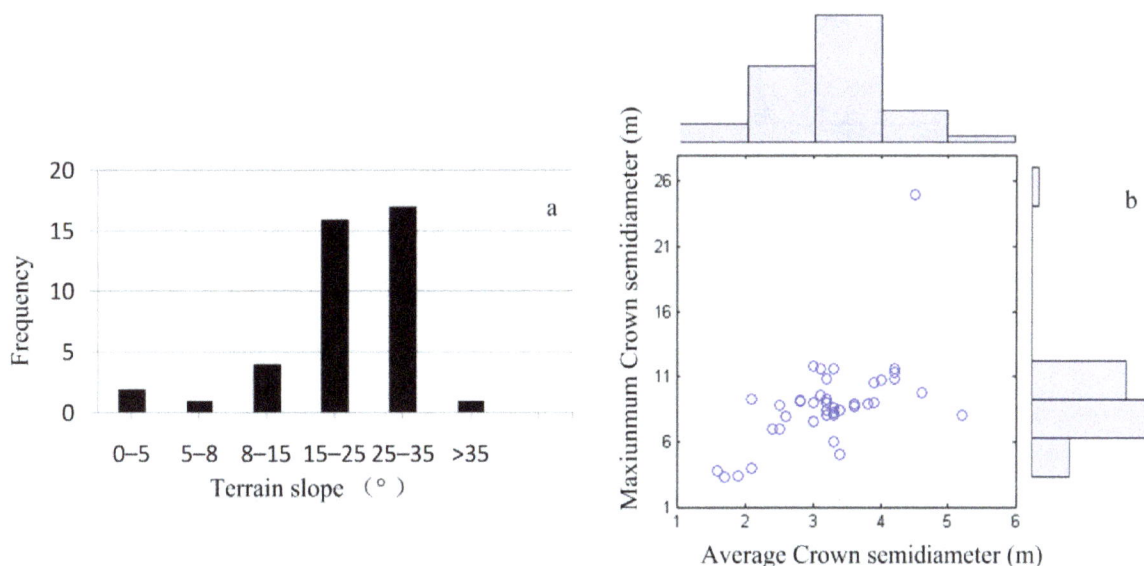

**Figure 2.** Distribution of terrain slope in the plots, average and maximum crown semidiameter. (**a**) terrain slope; (**b**) average and maximum crown semidiameter.

Plot number 34, a broadleaf forest plot with *S. superba*, *Mallotus paniculatus* and *Ficus variolosa* as the dominant tree species, is cited again as an example. This plot had average elevation 562.4 m and average slope 31.2°. Field surveys detected a total of 98 trees, with an average height of 7.86 m (maximum, 13.5 m; minimum, 3.6 m). The average, maximum, minimum DBH of all trees were 9.2 cm, 37.0 cm and 5.0 cm, respectively, average crown at 4.1 m (maximum, 16.8 m; minimum, 1.5 m). The plot point cloud comprised 14,047 points.

## 3. Methodology

### 3.1. Terrain Impacts on Canopy Height

Both the canopy height point cloud and CHM are by nature elevation differences between the point cloud data, DSM and DEM for the same point in a coordinate plane. For an upright tree, the canopy height should be the difference between the elevation of the canopy and the elevation where the roots enter the ground. However, when the ground surface slopes, which is often the case, or other complex terrain features are present such as ridges, valleys, escarpments, and eroded areas are present, the elevation differences deviate from the actual canopy height.

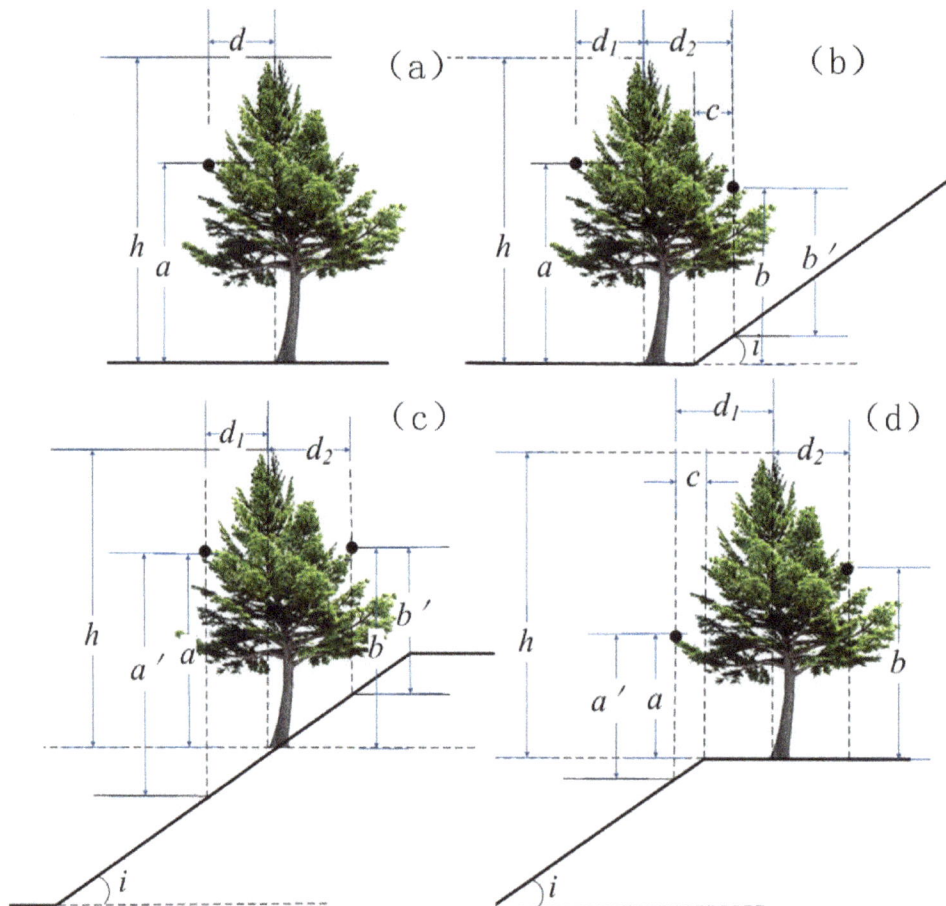

**Figure 3.** Influence of slope on canopy height.

Figure 3 shows the influence of ground slope on canopy height when the canopy height is calculated as the difference between point cloud data and the DEM. The black dots in the figure represent laser hits on the canopy, where $i$ is the angle of the slope gradient. On flat ground, the difference between the point cloud data and DEM reflects the actual canopy height (Figure 3a). The tree is located near the lower end of the slope, with part of its canopy covering the slope (Figure 3b). The difference between the elevation of the laser hit at the left side and DEM is its actual canopy height ($a$). The actual canopy height corresponding to the laser hit at the right side is $b$. However, the difference between the laser hit at the right side and the DEM leads to another height $b' = b - c \tan(i)$, which represents an error of $-c \tan(i)$. Similarly, when the tree stands on a slope (Figure 3c), canopy height calculated as the differences between DEM and laser hits at the left and the right side are $a' = a + d_1 \tan(i)$ and $b' = b - d_2 \tan(i)$, respectively, representing respective differences in $+d_1 \tan(i)$ and $-d_2 \tan(i)$. Additionally, where the tree is located at the top of the slope with its canopy covering part of the slope, the canopy height error at the left side is $+c \tan(i)$ (Figure 3d).

Terrain-induced canopy height differences are mainly determined by the slope gradient and the crown radius in that larger slope gradients and wider crowns lead to greater differences. When the laser hit is on the edge of the crown, and is in the upslope direction, the maximum height difference is expected to be observed. Laser hits farther away from the centre point of the trunk cause larger differences. The maximum differences of canopy height was calculated by the formula $d/2 \tan(i)$ (Table 2), where $d$ is the diameter of the crown, and $i$ is the slope gradient. Obviously, the influence of terrain on individual tree canopy height measurement is too significant to be ignored.

**Table 2.** Maximum canopy height difference calculated by slope gradient and crown.

| Slope $i$ (°) Crown $d$ (m) | 5 | 10 | 20 | 30 | 40 | 50 |
|---|---|---|---|---|---|---|
| 3 | 0.13 | 0.26 | 0.54 | 0.86 | 1.26 | 1.79 |
| 5 | 0.22 | 0.44 | 0.91 | 1.44 | 2.10 | 2.97 |
| 10 | 0.44 | 0.88 | 1.82 | 2.88 | 4.20 | 5.96 |
| 15 | 0.66 | 1.32 | 2.73 | 4.33 | 6.29 | 8.94 |

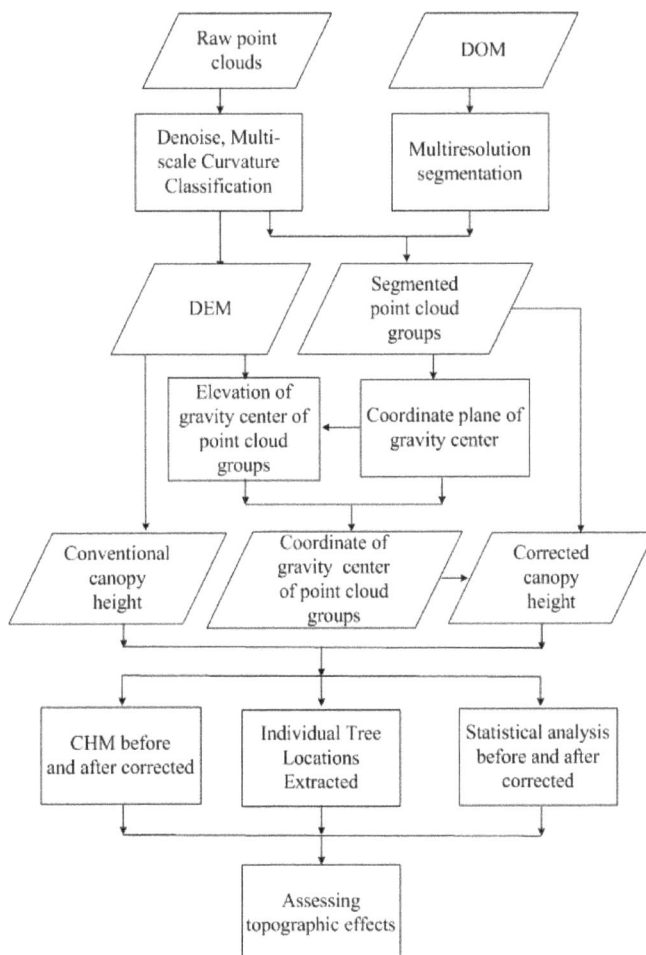

**Figure 4.** Flowchart of methods used in this study.

This shows that the terrain-induced canopy height difference is, in essence, the elevation difference between the laser hit and tree root elevation. That is, canopy height differences exist without exception because of the uneven ground. The differences will, in turn, influence the extraction of canopy vertices, trunk location, crown, volume and biomass. Extracted forest parameters will be distorted. It is thus imperative to remove the influence of terrain to conduct a more accurate extraction of canopy height. Because the normalized point cloud or CHM is obtained by subtracting the DEM data from DSM data or the raw point cloud, the influence of topography on canopy height varies with individual trees. Therefore, the point cloud of individual trees should be segmented first and foremost. Figure 4 shows a flowchart of the method used. The details are described below.

### 3.2. Processing of LiDAR Data

LiDAR points were classified into ground and non-ground (vegetation) returns with MCC. MCC was designed for classifying LiDAR returns in forested environments occurring in complex terrains [18]. The MCC algorithm operates by discarding returns that exceed a threshold curvature calculated using a surface interpolated [18]. Ranges for initial parameters were selected for MCC based on the scale and curvature ranging from 0.8 to 1.5 and 0.01–0.10, respectively. Ground returns were used to generate DEM intervals of 0.3m.

Studies have shown that first echo point cloud data can better reflect the forest canopy structure [24]. Therefore, the first echo point cloud was mostly employed in the current paper. Because the first echo point cloud higher than 1.8 m is believed to reflect the actual structure and morphology of the entire crown, these data are usually used for retrieval of forest parameters. Thus echo points of higher than 1.8 m above the ground were generally adopted as vegetation points to avoid the interference of shrub layer [25]. Moreover, point cloud percentiles could reflect the distribution of laser hits [26–28]. The proportions of points higher than given thresholds in the total number of points were calculated.

### 3.3. Crown Segmentation

To obtain individual tree height, the initial process is to isolate individual trees and delineate tree crown boundaries. Previous researchers have been done on isolating individual trees using large-scale aerial photos or high-spatial resolution remotely sensed imagery. The methods for isolating individual trees from imagery or photos include: edge detection using scale-space theory [29], local maxima detection [30], local maxima filtering with fixed or variable window sizes [31], local transect analysis [32], and watershed segmentation [33,34]. These methods are mostly based on the assumption that there are "peaks" of reflectance around the treetops and "valleys" along the canopy edges.

Crown segmentation was performed using the eCognition (Definiens Developer 8.7) software package, with multi-resolution segmentation (MS), and a digital orthophoto map (DOM) with a resolution of 0.2 m as input in our study. To avoid over-segmentation as a result of DOM, the segmentation process was applied to a median filtered version of DOM. The median filter applied to the images had a window size of $7 \times 7$ pixels. The MS algorithm required scale, shape ratio and Compactness ratio as input parameters, a scale parameter associated with the average size of resultant objects, a Shape ratio associated with the shape criterion of homogeneity, and a Compactness ratio associated with the optimization criteria for object shape. The parameters were adjusted as required for each image to account for differences in vegetation structure and distribution. The best segmentation parameters for different forest types were set after repeated experimentation. Segmentation was performed using different parameters: the scale parameter ranging from 10 to 14, the shape ratio ranged from 0.6 to 0.8 and the Compactness ratio ranged from 0.6 to 0.9. The final segments were reviewed manually to ensure quality.

### 3.4. Terrain Correction of Normalized Point Cloud

Point cloud data from individual trees were segmented based on crown boundaries which formed a closed polygon in the segmented DOM. The coordinate planes of centres of gravity were calculated through the weighted height of each point cloud group (Equation (1)). In Equation (1), $x_{ig}$, $y_{ig}$ are the coordinates of centres of gravity in the point cloud group $i$; $A_{ij}$ are the elevations of points in the point cloud group $i$ extracting from DEM; $x_{ij}$, $y_{ij}$, $z_{ij}$ are the 3D coordinates of point in the point cloud group $i$; $j$ is serial number of grounds $1, 2, ..., n$, while $n$ is the total number of points in group $i$:

$$
\begin{aligned}
x_{ig} &= \frac{\sum_{j=1}^{n} x_{ij} \cdot \left(z_{ij} - \frac{1}{n}\sum_{j=1}^{n} A_{ij}\right)}{\sum_{j=1}^{n} \left(z_{ij} - \frac{1}{n}\sum_{j=1}^{n} A_{ij}\right)} \\
y_{ig} &= \frac{\sum_{j=1}^{n} y_{ij} \cdot \left(z_{ij} - \frac{1}{n}\sum_{j=1}^{n} A_{ij}\right)}{\sum_{j=1}^{n} \left(z_{ij} - \frac{1}{n}\sum_{j=1}^{n} A_{ij}\right)}
\end{aligned}
\tag{1}
$$

For individual trees and forest gaps, the calculated coordinates represent the peak of the tree's canopy relative to the ground and the geometrical centre of the gap, respectively. When a tree stands upright, its peak is directly over the base of the tree. Because this paper focuses on the effects of terrain to canopy height and forest biomass, the forest gap point cloud was removed by setting a height threshold. Finally, elevation values of centres of gravity $z_{ig}$, namely the elevation values of the tree base, were extracted from DEM according to their same plane coordinates. Additionally, the difference between $z_{ig}$ and elevation of the point cloud in each group was considered to be the normalized point cloud after correction.

### 3.5. Individual Tree Locations Extracted

Before correction, CHM I was generated by subtracting DEM data from DSM data or the raw point cloud data. Normalized point cloud data based on individual tree crown segmentation was used to generate CHM II after correction through the Inverse Distance Weighted (IDW) interpolation method [35]. Individual tree crowns and individual tree locations were extracted to assess deviation caused by terrain of CHM retrieval from LiDAR data before and after correction. Individual tree crowns were also extracted by the canopy morphological-controlled watershed method from both CHMs before and after correction [36]. Morphological crown control was introduced to ensure that the watershed results are accurately located in the crown area. Additionally, both CHMs were used to extract individual tree locations through the region growing method [36]. The local maxima algorithm is used to identify potential tree positions in crown area.

Regression analysis was conducted using Microsoft Excel to assess the correlation between the slope gradient and the proportions of points with different thresholds of canopy height differences before and after correction.

### 3.6. Assessing the Consequence of Correcting Topographic Effects

We apply a mean stand height weighting scheme, named the Lorey's height (basal-area-weighted average height) [37], which already is in common use in forestry. Lorey's height is defined by Equation (2):

$$Lh = \frac{\sum G_i \cdot h_i}{\sum G_i} \tag{2}$$

where $Lh$ is the Lorey's height (basal-area-weighted average height), $G_i$ is the basal area of stem i, and $h_i$ is the height of stem i.

Stepwise multiple regressions were used to find a relationship between canopy heights variables and field surveyed Lorey's height. Canopy heights variables from the LiDAR data before and after terrain correction were used as independent variables in the regression analysis. Two models were built respectively (Equation (3)). We used a K-fold cross-validation procedure (JMP 10, SAS Institute, Cary, NC, USA) to identify the most appropriate dimension for the regression models. This procedure splits the dataset into K groups and fits a regression model to all groups except one. The model giving the best validation statistic is chosen as the final model. This method is best for small data sets, because it makes efficient use of limited amounts of data:

$$Lh = \beta_0 + \beta_1 h_{10} + \beta_2 h_{20} + \cdots + \beta_9 h_{90} + \beta_{10} h_{mean} + \varepsilon \tag{3}$$

where $h_{10}$, $h_{20}$, ..., $h_{90}$ and $h_{mean}$ is 10%, 20%, ..., 90% height quantile and average height of airborne LiDAR point cloud respectively; $\beta_0$, $\beta_1$, $\beta_2$, ..., $\beta_9$ and $\beta_{10}$ is the coefficient of model respectively; $\varepsilon$ is the error of model.

Stepwise variable selection and the maximum K-fold R-square improvement variable selection techniques were applied to select the LiDAR-derived variables to be included in the models [24]. The two Lorey's height estimation models based on 41 plots were named Models I and II, respectively. In assessing deviation caused by terrain of canopy height, we report $R^2$ for statistically significant (at $p < 0.001$) regressions, the RMSE, equation intercept and coefficients, and the maximum K-fold R-square.

## 4. Results

### 4.1. CHMs before and after Correction

Figure 5 shows the results of crown and point cloud analysis segmented by MS in plot No. 34. The MS segmentation parameters were set to the Scale parameter of 14, the Shape ratio is 0.6 and the Compactness is 0.8. The point cloud was separated into both individual tree and forest gap point clouds (Figure 5b).

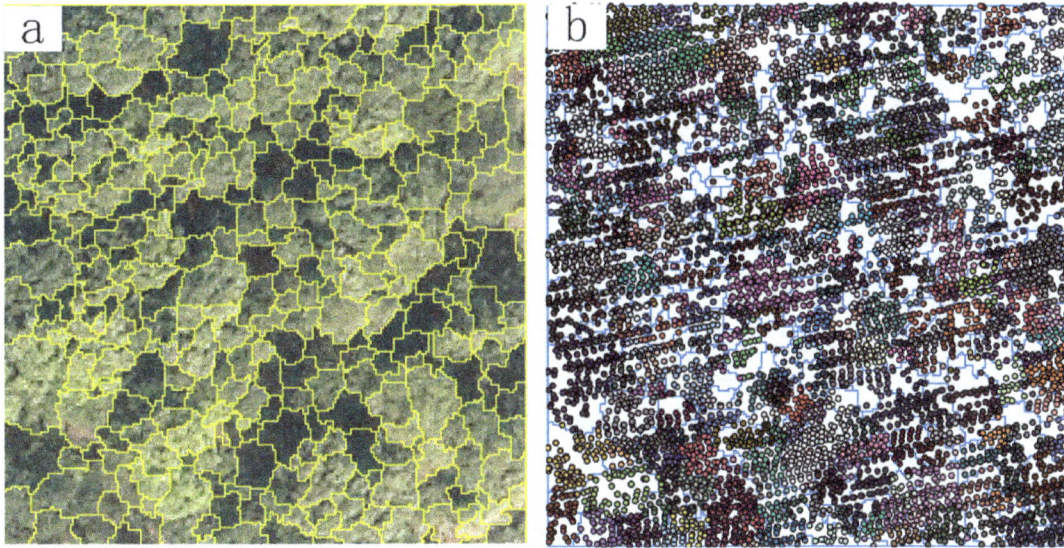

**Figure 5.** Multi-resolution segmentation in Plot No. 34: (**a**) segmented digital orthophoto map; (**b**) segmented point cloud.

Figure 6a,b shows CHM I and II based on the IDW interpolation before and after correction. The differences between CHMs before and after correction ranged from −2 m to 2 m (Figure 6c). The positive and negative values appear alternately along the direction of slope. The differences are negative at the upslope side from the centre point of the trunk, and are positive at the downslope side.

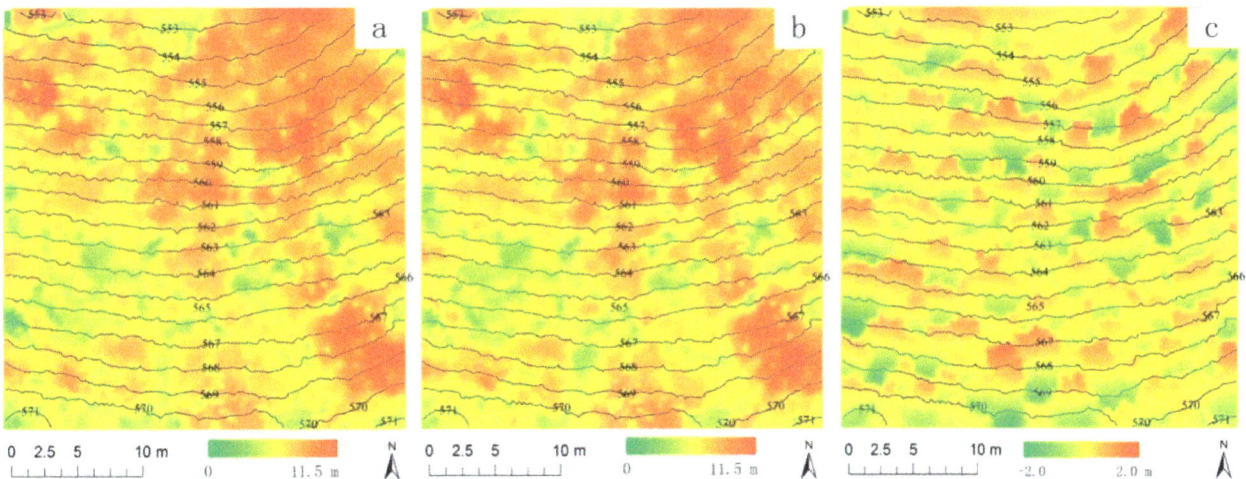

**Figure 6.** Canopy model heights (CHMs) before and after correction and difference image in plot No. 34: (**a**) CHM I before correction (**b**) CHM II after correction; (**c**) image of the differences.

*4.2. Impact of Topography on Individual Tree Extraction*

Figure 7 shows the point cloud of an individual tree in the study area. The ground slope gradient is 38.2°, tree height is 24.4 m, and crown radius is 6.5 m and 5.5 m in cross direction. Figure 7a shows the raw point cloud data; Figure 7b presents the new normalized point cloud proposed in this paper; Figure 7c shows normalized point cloud data by the conventional method, *i.e.*, by subtracting the DEM from the raw point cloud. The long dashed lines represent the trunk. Point cloud data in Figure 7a,b are highly consistent with tree morphology, while Figure 7a,c are quite different. To be specific, the point cloud data in the oval box on the left side of the dashed line of Figure 7c are higher than the raw point cloud, whereas those in the square box on the right side of the dashed line are lower than the raw point cloud. The left and right sides of the dashed line is the downslope and upslope sides, respectively. In other words, the conventional normalized point cloud, which is the differences between raw point cloud and DEM, actually increases the modelled downslope side of the canopy by up to 2.23 m; while the upslope side of the canopy is lowered by up to −2.34 m. The extent to which the canopy height is raised or lowered is related to the slope gradient and the horizontal distance between the point cloud and the trunk. Greater gradients and longer distances from the point cloud to the trunk cause greater height differences. Points close to the trunk centre display nearly zero difference.

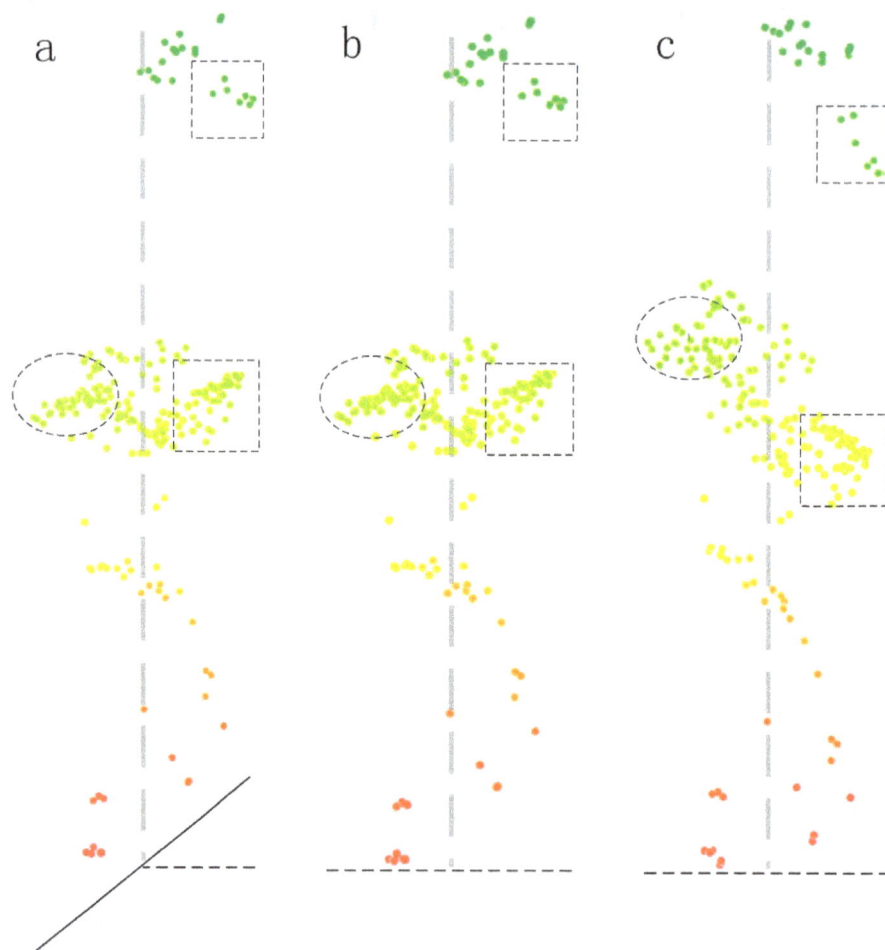

**Figure 7.** Point cloud of individual tree canopies: (**a**) raw point cloud; (**b**) new normalized point cloud; (**c**) normalized point cloud by a conventional method.

CHM is a key factor affecting the accuracy of parameter extraction from individual trees in that CHM directly influences the extracted location and crown of single trees. CHM is acquired from the normalized point cloud after interpolation. Figure 8 shows individual tree crowns and individual tree locations extracted. The results showed differences in both the number of trees and tree positions: 62 trees from CHM I and 63 from CHM II and a maximum distance difference of 1.57 m in terms of tree location (Figure 8a). The number of crowns based on CHM I and CHM II were 71 and 68, respectively, representing some integrations and separations. Apart from different crown numbers, there was also a considerable amount of difference in the shape and size of crowns (Figure 8b,c).

**Figure 8.** Individual tree location and crown comparison: CHM I *vs.* CHM II. (**a**) Individual tree locations extracted from CHM I and CHM II; (**b**) crowns extracted from individual trees based on CHM I; (**c**) crowns extracted from individual trees based on CHM II.

Because the traditional normalization method results in raising or lowering a point cloud, the normalized point cloud shows a much different morphology from the raw point cloud data, making it impossible to capture the accurate canopy structure of trees. There would be significant variation in the canopy vertices, individual tree locations and crowns extracted using these two methods.

*4.3. Impact of Topography on Plot-Level Canopy Height*

Point clouds in plot-level data areas are often used to establish biomass estimation models. Normalized point clouds and related statistics at the plot level have important effects on the precision of forest biomass models making it important to analyse how topography impacts canopy height at the plot level so that the precision of biomass estimation can be improved.

Figure 9a–c shows canopy heights in plot No. 34 before and after correction in ascending order as well as their differences. Figure 9d presents canopy height differences in ascending order; canopy height differences before and after correction, *i.e.*, terrain-triggered canopy height differences, are within ±2.0 m, and show a symmetrical distribution (Figure 9). This is the main reason that the ground slopes are basically uniform, and canopy has a symmetrical form with the trunk as the axis. The normalized point cloud at the downslope side was elevated with the normalized point cloud by the conventional method, so the differences are positive. Conversely, the normalized point cloud at the upslope side was lowered in elevation, so their differences are negative. In Figure 9b, with a canopy height showing the new normalized point cloud data of within ±1.8 m that corresponds to shrub or ground points, data within ±1.8 m is excluded by setting a threshold when forest parameters are extracted, such as tree height, crown, height to the first live branch, canopy density, biomass and so on. In other words, canopy height of less than ±1.8 m has nothing to do with forest parameters. The remaining point cloud whose canopy height is taller than 1.8 m after correction reflects the natural form of the canopy structure, and is thus conducive to subsequent extraction of parameters as listed above.

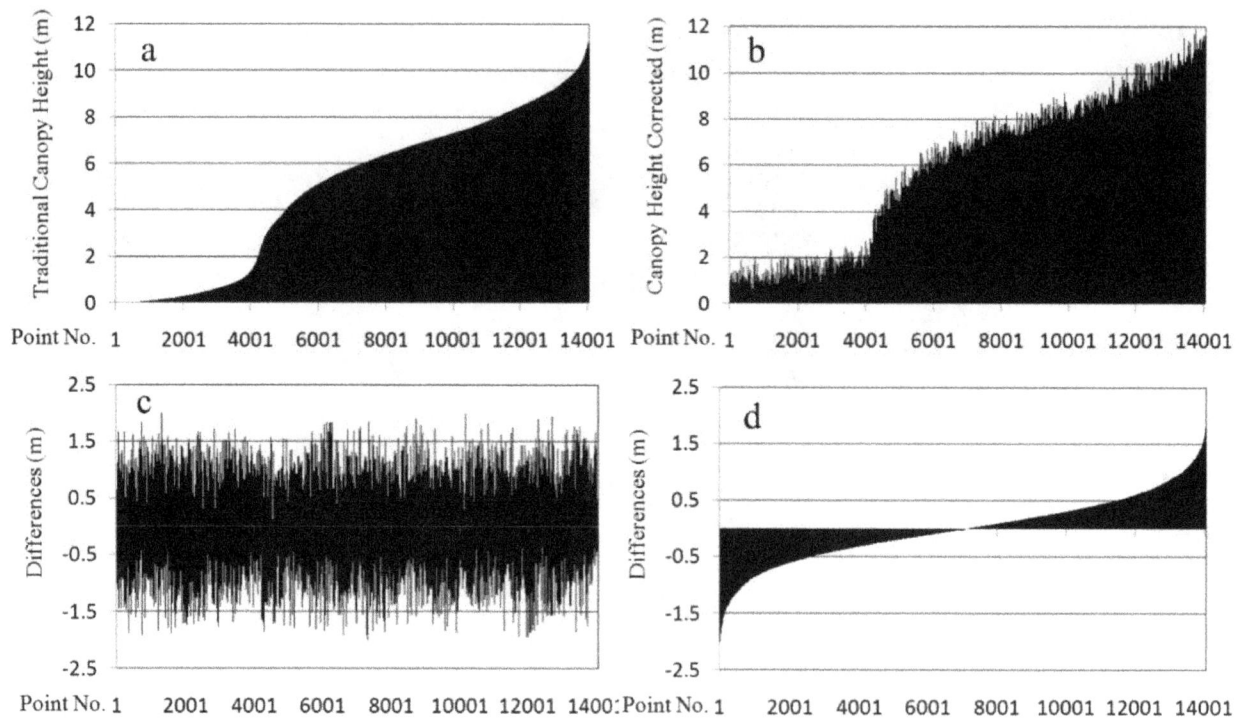

**Figure 9.** Canopy height before and after correction and canopy height differences in plot No. 34: (**a**) canopy height before correction; (**b**) canopy height after correction; (**c**) canopy height differences before and after correction; (**d**) canopy height differences in ascending order.

The absolute value of terrain-caused canopy height differences of all plots was calculated and analysed. A difference threshold, $k$, was set at 0.3, 0.5, 0.8, 1.0, 1.2, and 1.5 m for various tests. Average values of differences above each threshold $k$ are labelled as $mean_{0.3}$, $mean_{0.5}$, $mean_{0.8}$, $mean_{1.0}$, $mean_{1.2}$, and $mean_{1.5}$, respectively. The proportions of the point counts of canopy height differences higher than a given threshold $k$ in the total number of points are labelled as $p_{0.3}$, $p_{0.5}$, $p_{0.8}$, $p_{1.0}$, $p_{1.2}$, and $p_{1.5}$, respectively. The plots were numbered based on their slope gradients in ascending order.

Figure 10 shows average differences before and after canopy height correction. A zero average difference indicates no difference is greater than the corresponding threshold $k$. Based on the statistical results, the average values of $mean_{0.3}$, $mean_{0.5}$, $mean_{0.8}$, $mean_{1.0}$, $mean_{1.2}$, and $mean_{1.5}$ of all plots vary: 0.58, 0.72, 0.97, 1.11, 1.21, and 1.28 m, respectively, and the maximum values of all plots were 0.83, 0.98, 1.25, 1.43, 1.67, and 1.95 m, respectively. In general, the average value of differences increases as the slope becomes steeper (Figure 10). However, when the slope is $\geq 22°$, the average canopy height difference barely increases. This can be mostly explained by the small crown radius within the plots which is average 3.5 m. Smaller crowns indicate that slope has less impact on the canopy height than larger crowns. Additionally, the canopy height differences tend to be symmetrical, which means that similar characteristics are observed regardless of whether the differences are negative or positive.

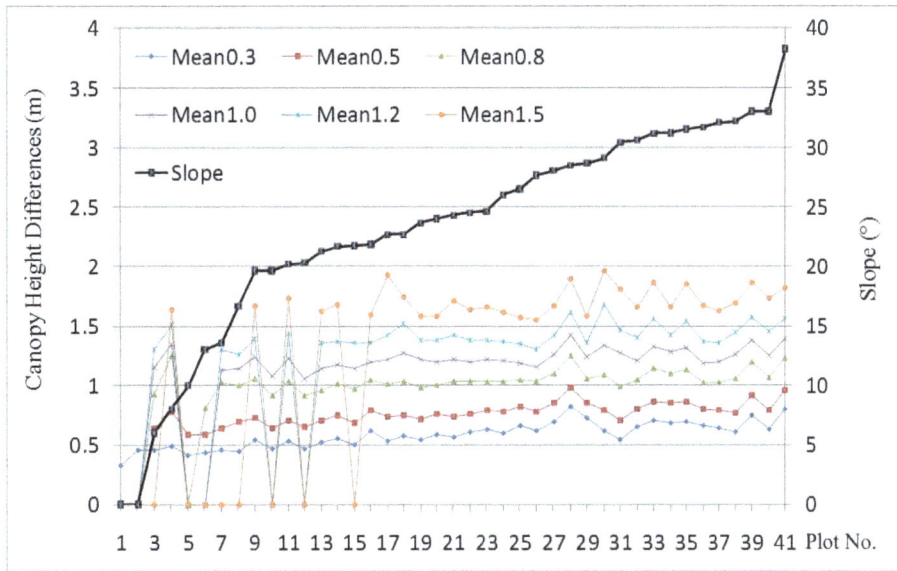

**Figure 10.** Average values of terrain-triggered canopy height differences *versus* slope.

Figure 11 shows the proportions of points higher than given thresholds for the total number of points. Average values of $p_{0.3}$, $p_{0.5}$, $p_{0.8}$, $p_{1.0}$, $p_{1.2}$, and $p_{1.5}$ of all plots were 20.0%, 11.0%, 4.2%, 2.2%, 1.1%, and 0.35%, respectively, and maximum values of $p_{0.3}$, $p_{0.5}$, $p_{0.8}$, $p_{1.0}$, $p_{1.2}$, and $p_{1.5}$ of all plots were 32.6%, 23.2%, 12.9%, 8.5%, 5.3%, and 2.6%, respectively. When the slope increases from zero to 38.2°, the proportions of points with differences larger than 0.3, 0.5, 0.8, 1.0, 1.2 and 1.5 m increase from 0% to 32.6%, 23.2%, 12.9%, 8.5%, 5.3% and 2.6%, respectively (Figure 11). Overall, the proportion increases as the slope increases, but it decreases as the threshold grows.

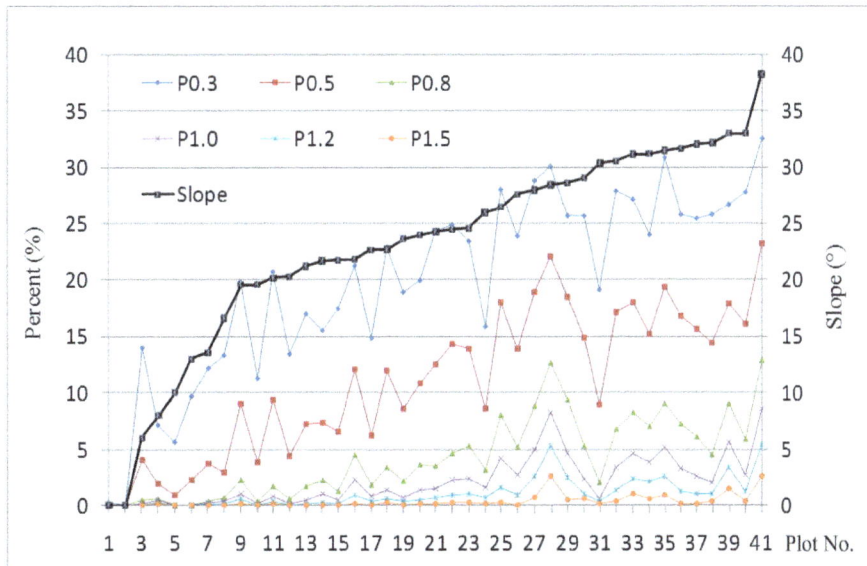

**Figure 11.** Proportions of points with differences greater than given thresholds *versus* slope.

Meanwhile, regression analysis shows a strong correlation between the slope gradient and the proportions of points with differences greater than 0.3, 0.5 and 0.8 m (Figure 12); the coefficient of determination $R^2$ is 0.83, 0.76, and 0.60 (n = 41), respectively. But when the threshold is increased to

≥1.0 m, the correlation weakens. This is also attributable to smaller crowns, which means terrain creates less impact. Larger crowns apparently lead to more slope-caused canopy height differences.

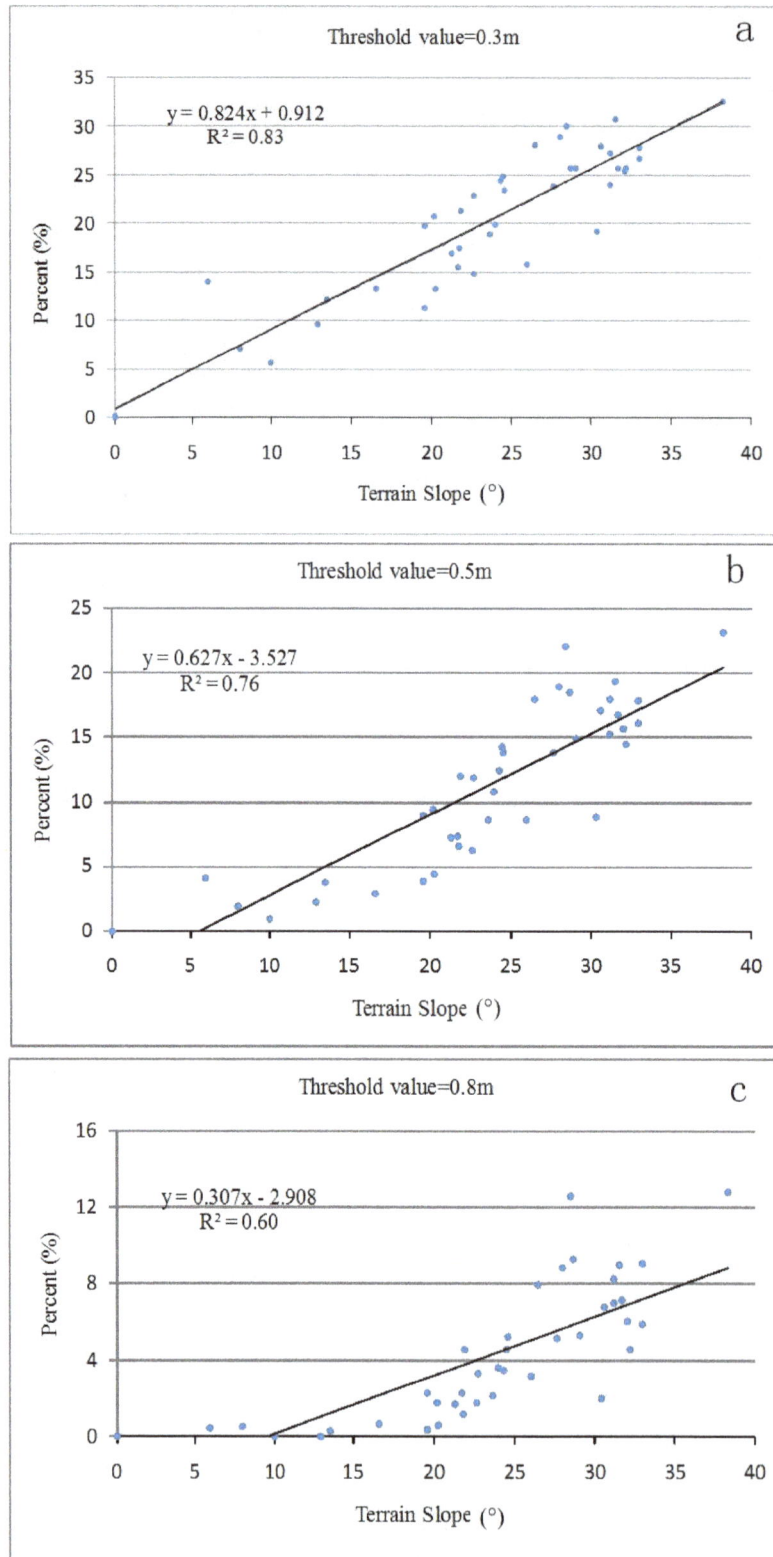

**Figure 12.** Regression relationships between slope and proportions of points with differences greater than selected thresholds: (**a**) threshold = 0.3 m; (**b**) threshold = 0.5 m; (**c**) threshold = 0.8 m.

## 4.4. Consequence of Correcting Topographic Effects

To assess the consequence of correcting topographic effects on forest canopy height, the two best regression models were chosen and built according to the maximum R-square stopping rule from 10-fold cross validation, namely Model I and Model II (Table 3). The height of the canopy at the 10th, 40th, 60th, 90th percentile and average height were selected in Model I, the determination coefficient $R^2$ is 0.61, corresponding adjusted $R^2$ is 0.58, $RMSE$ is 2.24 m, the response mean is 12.72 m, and the maximum K-fold $R^2$ is 0.52. The height of the canopy at the 10th, 40th, 50th, 70th, 80th, 90th percentile and average height were selected in Model II, the determination coefficient $R^2$ is 0.77, corresponding adjusted $R^2$ is 0.71, $RMSE$ is 1.86 m, and the response mean is 12.72 m, and the maximum K-fold $R^2$ is 0.62 respectively (Figure 13). Experimental results show that the correlation between Lorey's height calculated by filed survey and canopy height quantiles after terrain correcting is better than before terrain correcting, which reveals that normalized canopy height point cloud after the terrain correction is closer to the natural formation of forest canopy. It can demonstrate that the method of terrain correction restores natural forms of forest canopy.

**Table 3.** Regression coefficients, estimated values and precision indexes of the models.

| Coefficients | Model I ( n = 41) | | | | Model II ( n = 41) | | | |
|---|---|---|---|---|---|---|---|---|
| | Estimated Values | Error Sum of Squares (SS) | F Ratio | P > F | Estimated Values | Error Sum of Squares (SS) | F Ratio | P > F |
| $\beta_0$ | 1.382009 | 0.0 | 0.000 | 1 | 0.814054 | 0.0 | 0.000 | 1 |
| $\beta_1$ | −3.0262 | 36.37898 | 8.80696 | 0.005382 | −3.33578 | 34.44685 | 9.895305 | 0.003499 |
| $\beta_2$ | | | | | | | | |
| $\beta_3$ | | | | | | | | |
| $\beta_4$ | 2.62023n7 | 34.95306 | 8.46176 | 0.006263 | 3.454034 | 26.42372 | 7.590555 | 0.009475 |
| $\beta_5$ | | | | | −6.26901 | 21.0267 | 6.04019 | 0.019406 |
| $\beta_6$ | −3.33947 | 40.49818 | 9.804176 | 0.003506 | | | | |
| $\beta_7$ | | | | | 8.454889 | 19.45856 | 5.589724 | 0.024103 |
| $\beta_8$ | | | | | −7.81642 | 29.14998 | 8.37371 | 0.006697 |
| $\beta_9$ | 0.569135 | 23.73004 | 5.744788 | 0.022012 | 1.141369 | 52.70873 | 15.14126 | 0.000458 |
| $\beta_{10}$ | 3.614737 | 19.06262 | 4.614855 | 0.038697 | 5.001009 | 25.61142 | 7.357213 | 0.010527 |

## 5. Discussion

This paper offers a correction method of terrain-induced canopy height differences under the premise that the trees stand vertically. Of course, the suggested correction may not be applicable to tilting trees. Accurate measurement of the canopy height of tilting trees is more complicated, although this can be performed using auxiliary ground measurements [7].

Another premise of this paper is that the point clouds representing individual trees must be generated by DOM segmentation. The segmentation process relies on user-specified parameters regarding the scale, shape ratio and Compactness ratio in this study. When segmenting crown from remote sensing image and aerial photography image with high resolution, these methods are mostly based on spectral properties. However, the "peaks" and "valleys" are not always distinct since canopy reflectance is affected by various factors such as illumination conditions, canopy spectral properties, and complex canopy structure.

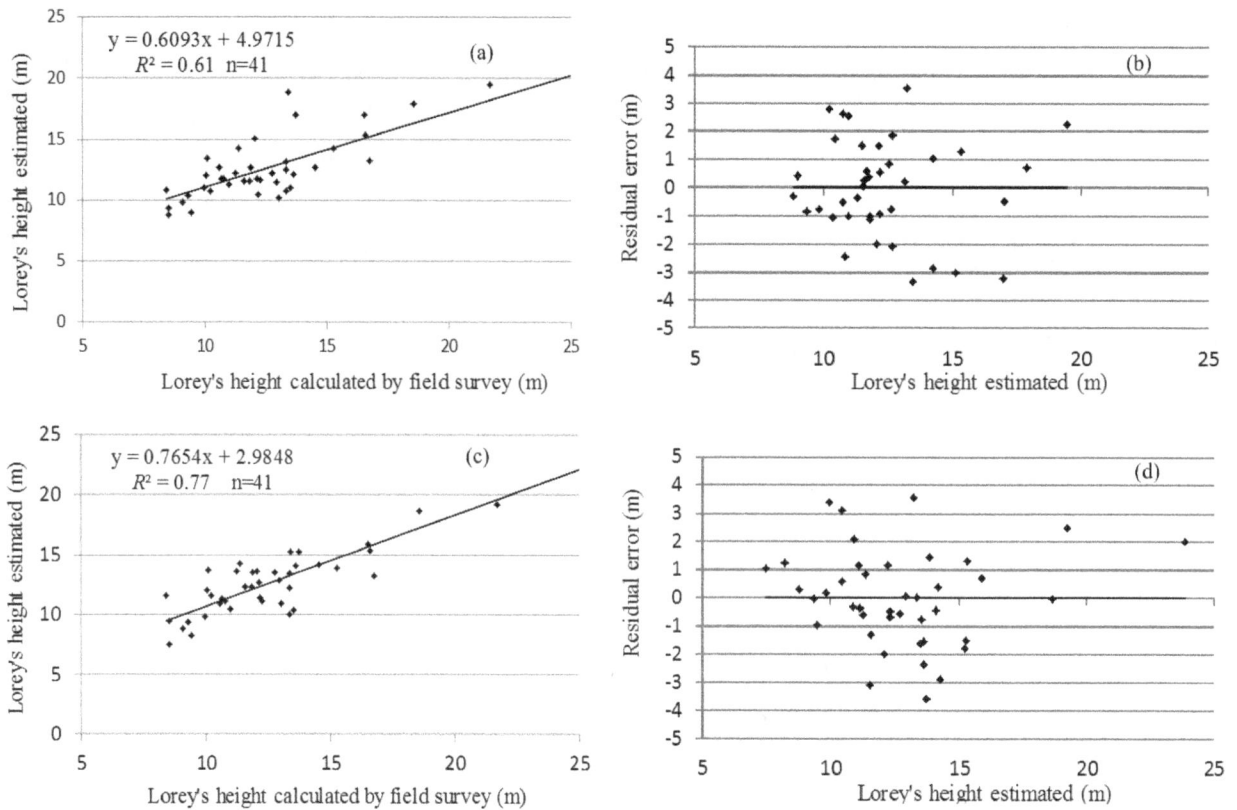

**Figure 13.** Correlation and residual error of the estimated Lorey's height before and after terrain correction. (**a**) Correlation of the estimated Lorey's height before terrain correction; (**b**) Residual error of the estimated Lorey's height before terrain correction; (**c**) Correlation of the estimated Lorey's height after terrain correction; (**d**) Residual error of the estimated Lorey's height after terrain correction.

Meanwhile, there may be overlapping point clouds in complex terrain and/or closed-canopy forest. Moreover, because the airborne LiDAR scanning was done at a large side-viewing angle, shadows fall into the DOM image. Therefore, only point clouds within the plots were segmented in this study and there may be some errors in the segmentation. Nevertheless, the DOM segmentation was done after repeated experimentation, and we believe that point clouds of single trees obtained this way had only minor errors. Segmentation of point clouds for the entire region is not currently feasible and warrants further study.

The crown segmentation is the key procedure for terrain correction of forest canopy height retrieval using airborne LiDAR. Another crown segmentation method bases on airborne LiDAR data. Compared with passive imaging, LiDAR has the ability of directly measuring the three-dimensional coordinates of canopies. Therefore, the geometric, rather than spectral, "peaks" and "valleys" can be detected. Researchers have applied LiDAR data into crown segmentation and individual tree isolation directly [38–40]. The crown segmentation methods using airborne LiDAR can be divided into two kinds, and they are based on discrete point cloud and with surfaces derived from point cloud. The first method segments crown using the point cloud directly. The later method usually transforms the point cloud into a raster. The first and final returns were used for generating the DSM and the CHM. The CHM has inherent errors and uncertainties from a number of sources. Such LiDAR-derived surface models often contain so-called "pits" which occur. The pit-free algorithm can be used for generating the pit-free CHM [41]. Meanwhile, spatial error can be introduced during the interpolation process from the point cloud to the gridded height model [7]. The uncertainty introduced by the interpolation method can affect the accuracies of the segmentations. Compared with using CHM, the crown segmentation methods using discrete point cloud can take advantage of the full 3D structure inherent in the LiDAR point cloud. Many approaches using point clouds have been developed to segment crown, such as region growing [42], marker-controlled watershed segmentation [43], combination of pouring algorithm and knowledge-based assumptions [44], adaptive clustering [45]. Jakubowski *et al.* [46] compared a 3D lidar point cloud segmentation algorithm to an object-based image analysis (OBIA) of CHM to determine the difference between the two types of approaches. The two approaches delineated tree boundaries that differed in shape: the lidar-approach produced fewer, more complex, and larger polygons that more closely resembled real forest structure. Effectively, the lidar segmentation method tended to under-segment and under-detect trees, while the OBIA method over-segmented the trees. However, there are over-segmentation and under-segmentation in segmenting forest crown using remote sensing image and using airborne LIDAR data. Further research is necessary to improve precision of segmentation and to automate the segmentation process. However, the correction method based on individual tree crown segmentation is only applicable on the plot scale, because of overlapping point clouds caused by complex terrain, closed forest canopies, and other reasons. Consequently, the segmentation of point clouds representing the entire region is worth further study.

Geoscience Laser Altimeter System (GLAS, footprint size = ~70 m), with a larger-footprint and wide spatial coverage, has provided practical means for monitoring various global forest attributes. The sloped terrain generally lengthens the full extent of GLAS waveform and decreases the level of laser energy at the forest canopy and ground peaks [47]. Lee *et al.* [48] found that the larger footprint size and greater slope tend to generate more errors in the retrieved lidar forest canopy heights. Park *et al.* [49] found that both Laser Vegetation Imaging Sensor (LVIS, footprint size = ~20 m) and GLAS campaigns could be benefited from the physical correction approach, and the magnitude of accuracy improvement was determined by footprint size and terrain slope, off-nadir pointing angle. Our results indicate that airborne LiDAR data also can be benefited from topographic correction. It can be concluded that both large and small footprint LiDAR data will encounter high terrain slop problem, and topographic correction is required.

## 6. Conclusions

Canopy height serves as basic data for the extraction of forest parameters using LiDAR. However, canopy height obtained by subtracting DEM data from DSM data contains errors because of the influence of terrain. This paper proposes a method for correcting terrain-derived canopy height differences based on individual tree crown segmentation. The new normalized point clouds are very consistent with raw point clouds morphologically. The method can obtain accurate normalized canopy heights which recover the natural structure of the canopy.

The results show canopy height differences are connected with slope gradients and crown radius. Steeper slopes and larger crowns cause greater differences. For individual trees, terrain influences the estimated canopy height of individual trees when the DEM is subtracted from the DSM, the downslope side of the tree trunk is elevated and the upslope side is lowered. For plot scale measurement, a strong correlation exists between the slope gradient and the proportions of points with differences higher than 0.3, 0.5 and 0.8 m, with the coefficient of determination $R^2$ at 0.83, 0.76, and 0.60 (n = 41), respectively.

## Acknowledgments

We acknowledge financial support from the National Natural Science Foundation of China (No. 41201351) and (No. 41401508), also the Strategic Priority Research Program—Climate Change: Carbon Budget and Related Issues of the Chinese Academy of Sciences (Grant No. XDA05050108). We also thank the support from the South China Botanical Garden, Chinese Academy of Sciences for the field work.

## Author Contributions

Zhugeng Duan carried out the analyses and wrote the manuscript. Dan Zhao and Yujin Zhao were responsible for recruitment the participants, and design and follow-up the experiments. All the authors were responsible for data cleaning. All the authors drafted the manuscript, and approved the final manuscript.

## Conflicts of Interest

The authors declare no conflict of interest.

## References

1. Falkowski, M.J.; Wulder, M.A.; White, J.C.; Gillis, M.D. Supporting large-area, sample-based forest inventories with very high spatial resolution satellite imagery. *Prog. Phys. Geogr.* **2009**, *33*, 403–423.
2. Maltamo, M.; Packalén, P.; Yu, X.; Eerikäinen, K.; Hyyppä, J.; Pitkänen, J. Identifying and quantifying structural characteristics of heterogeneous boreal forests using laser scanner data. *Forest Ecol. Manag.* **2005**, *216*, 41–50.
3. Næsset, E.; Bjerknes, K.O. Estimating tree heights and number of stems in young forest stands using airborne laser scanner data. *Remote Sens. Environ.* **2001**, *78*, 328–340.
4. Evans, J.S.; Hudak, A.T.; Faux, R.; Smith, A. Discrete return LiDAR in natural resources: recommendations for project planning, data processing, and deliverables. *Remote Sens.* **2009**, *1*, 776–794.
5. Takahashi, T.; Yamamoto, K.; Senda, Y.; Tsuzuku, M. Estimating individual tree heights of sugi (*Cryptomeria japonica* D. Don) plantations in mountainous areas using small-footprint airborne LiDAR. *J. Forest Res.* **2005**, *10*, 135–142.
6. Hodgson, M.E.; Bresnahan, E. Accuracy of airborne LiDAR-derived elevation: Empirical assessment and error budget. *Photogramm. Eng. Remote Sens.* **2004**, *70*, 331–339.

7.  Gatziolis, D.; Fried, J.S.; Monleon, V.S. Challenges to estimating tree height via LiDAR in closed-canopy forests: a parable from western Oregon. *Forest Sci.* **2010**, *56*, 139–155.

8.  Breidenbach, J.; Koch, B.; Kändler, G.; Kleusberg, A. Quantifying the influence of slope, aspect, crown shape and stem density on the estimation of tree height at plot level using LiDAR and InSAR data. *Int. J. Remote Sens.* **2008**, *29*, 1511–1536.

9.  Gousie, M.B.; Franklin, W.R. Augmenting grid-based contours to improve thin plate DEM generation. *Photogramm. Eng. Remote Sens.* **2005**, *71*, 69–79.

10. Guo, Q.; Li, W.; Yu, H.; Alvarez, O. Effects of topographic variability and LiDAR sampling density on several DEM interpolation methods. *Photogramm. Eng. Remote Sens.* **2010**, *76*, 701–712.

11. Kraus, K.; Pfeifer, N. Determination of terrain models in wooded areas with airborne laser scanner data. *ISPRS J. Photogramm.* **1998**, *53*, 193–203.

12. Elmqvist, M.; Jungert, E.; Lantz, F.; Persson, A.; Soderman, U. Terrain modelling and analysis using laser scanner data. *ISPRS Arch.* **2001**, *34*, 219–226.

13. Kobler, A.; Pfeifer, N.; Ogrinc, P.; Todorovski, L.; Oštir, K.; Džeroski, S. Repetitive interpolation: A robust algorithm for DTM generation from Aerial Laser Scanner Data in forested terrain. *Remote Sens. Environ.* **2007**, *108*, 9–23.

14. Axelsson, P. Processing of laser scanner data—Algorithms and applications. *ISPRS J. Photogramm.* **1991**, *54*, 138–147.

15. Vosselman, G. Slope based filtering of laser altimetry data. *ISPRS Arch.* **2000**, *33*, 935–942.

16. Sithole, G. Filtering of laser altimetry data using a slope adaptive filter. *ISPRS Arch.* **2001**, *34*, 203–210.

17. Chen, Q. Airborne LiDAR data processing and information extraction. *Photogramm. Eng. Remote Sens.* **2007**, *73*, 109–112.

18. Evans, J.S; Hudak, A.T. A multiscale curvature algorithm for classifying discrete return LiDAR in forested environments. *IEEE. Trans. Geosci. Remote Sens.* **2007**, *45*, 1029–1038.

19. Popescu, S.C. Estimating biomass of individual pine trees using airborne LiDAR. *Biomass Bioenergy* **2007**, *31*, 646–655.

20. Zeng, Y.; Schaepman, M.E.; Wu, B.; Clevers, J.G.P.W.; Bregt, A.K. Scaling-based forest structural change detection using an inverted geometric-optical model in the three gorges region of china. *Remote Sens. Environ.* **2008**, *112*, 4261–4271.

21. Leckie, D.; Gougeon. F.; Hill, D.; Quinn, R.; Armstrong, L.; Shreenan, R. Combined high-density LiDAR and multispectral imagery for individual tree crown analysis. *Can. J. Remote Sens.* **2003**, *29*, 633–649.

22. Lim, K.; Treitz, P.; Baldwin, K.; Morrison, I.; Green, J. LiDAR remote sensing of biophysical properties of tolerant northern hardwood forests. *Can. J. Remote Sens.* **2003**, *29*, 658–678.

23. Koukoulas, S.; Blackburn, G.A. Quantifying the spatial properties of forest canopy gaps using LiDAR imagery and GIS. *Int. J. Remote Sens.* **2004**, *25*, 3049–3072.

24. Næsset, E.; Gobakken, T. Estimation of above-and below-ground biomass across regions of the boreal forest zone using airborne laser. *Remote Sens. Environ.* **2008**, *112*, 3079–3090.

25. Nilsson, M. Estimation of tree heights and stand volume using an airborne LiDAR system. *Remote Sens. Environ.* **1996**, *56*, 1–7.

26. Ferster, C.J.; Coops, N.C., Trofymow, J.A. Aboveground large tree mass estimation in a coastal forest in British Columbia using plot-level metrics and individual tree detection from LiDAR. *Can. J. Remote Sens.* **2009**, *35*, 270–275.

27. Lefsky, M.A.; Cohen, W.B.; Acker, S.A.; Parker, G.G.; Spies, T.A.; Harding, D. LiDAR remote sensing of the canopy structure and biophysical properties of Douglas-fir western hemlock forests. *Remote Sens. Environ.* **1999**, *70*, 339–361.

28. Næsset, E. Predicting forest stand characteristics with airborne scanning laser using a practical two-stage procedure and field data. *Remote Sens. Environ.* **2002**, *80*, 88–99.

29. Brandtberg, T.; Walter, F. Automated delineation of individual tree crowns in high spatial resolution aerial images by multiple-scale analysis. *Mach. Vis. Appl.* **1998**, *11*, 64–73.

30. Dralle, K.; Rudemo, M. Stem number estimation by kernel smoothing of aerial photos. *Can. J. Remote Sens.* **1996**, *26*, 1228–1236.

31. Wulder, M.; Niemann, K.O.; Goodenough, D. Local maximum filtering for the extraction of tree locations and basal area from high spatial resolution imagery. *Remote Sens. Environ.* **2000**, *73*, 103–114.

32. Pouliot, D.A.; King D.J.; Bell, F.W.; Pitt, D.G. Automated tree crown detection and delineation in high-resolution digital camera imagery of coniferous forest regeneration. *Remote Sens. Environ.* **2002**, *82*, 322–334.

33. Schardt, M., Ziegler, M.; Wimmer, A.; Wack, R.; Hyyppae, J. Assessment of Forest Parameters by Means of Laser Scanning. *ISPRS Arch.* **2002**, 302–309.

34. Wang, L.; Gong, P.; Biging, G.S. Individual Tree-Crown Delineation and Treetop Detection in High-Spatial-Resolution Aerial Imagery. *Photogramm. Eng. Remote Sens.* **2004**, *70*, 351–358.

35. Kravchenko, A.; Bullock, D.G. A comparative study of interpolation methods for mapping soil properties. *Agron. J.* **1999**, *91*, 393–400.

36. Zhao, D.; Pang, Y.; Li, Z.Y.; Sun, G.Q. Filling invalid values in a LiDAR-derived canopy height model with morphological crown control. *Int. J. Remote Sens.* **2013**, *34*, 4636–4654.

37. Avery, T.E; Burkhart, H. *Forest Measurments*, 5th ed. McGraw-Hill: New York, NY, USA, 2002.

38. Hyyppä, J.; Kelle, O.; Lehikoinen, M.; Inkinen, M. A segmentation-based method to retrieve stem volume estimates from 3-D tree height models produced by laser scanners. *IEEE. Trans. Geosci. Remote Sens.* **2001**, *39*, 969–975.

39. Persson, Å.; Holmgren, J.; Söderman, U. Detecting and measuring individual trees using an airborne laser scanner. *Photogramm. Eng. Remote Sens.* **2002**, *68*, 925–932.

40. Brandtberg, T.; Warner, T.A.; Landenberger, R.E.; McGraw, J.B. Detection and analysis of individual leaf-off tree crowns in small footprint, high sampling density Lidar data from eastern deciduous forest in North America. *Remote Sens. Environ.* **2003**, *85*, 290–303.

41. Duan, Z.G.; Zeng, Y.; Zhao, D.; Wu, B.F.; Zhao, Y.J. Zhu, J.J. Method of removing pits of canopy height model from airborne LiDAR. *Trans. CSAE* **2014**, *30*, 209–217. (in Chinese with English abstract)

42. Chen, Q.; Baldocchi, D.; Gong, P.; Kelly, M. Isolating individual trees in a savanna woodland using small footprint LiDAR data. *Photogramm. Eng. Remote Sens.* **2006**, *72*, 923–932.

43. Tiede, D.; Hochleitner, G.; Blaschke, T. A full GIS-based workflow for tree identification and tree crown delineation using laser scanning, CMRT, 2005, Vienna, Austria, 29–30 August 2005; pp. 9–14.

44. Koch, B.; Heyder, U.; Weinacker, H. Detection of Individual Tree Crowns in Airborne LiDAR Data. *Photogramm. Eng. Remote Sens.* **2006**, *72*, 357–363.

45. Lee, H.; Slatton, K.C.; Roth, B.E.; Cropper JR, W.P. Adaptive clustering of airborne LiDAR data to segment individual tree crowns in managed pine forests. *Int. J. Remote Sens.* **2010**, *31*, 117–139.

46. Jakubowski, M.K.; Li, W.K.; Guo, Q.H.; Kelly, M. Delineating Individual Trees from LiDAR Data: A Comparison of Vector- and Raster-based Segmentation Approaches. *Remote Sens.* **2013**, *5*, 4163–4186.

47. Yang, W.; Ni-Meister, W.; Lee, S. Assessment of the impacts of surface topography, off-nadir pointing and vegetation structure on vegetation LiDAR waveforms using an extended geometric optical and radiative transfer model. *Remote Sens. Environ.* **2011**, *115*, 2810–2822.

48. Lee, S.; Ni-Meister, W.; Yang, W.Z.; Chen, Q. Physically based vertical vegetation structure retrieval from ICESat data: Validation using LVIS in White Mountain National Forest, New Hampshire, USA. *Remote Sens. Environ.* **2011**, *115*, 2776–2785.

49. Park, T.; Kennedy, R.E.; Choi, S.; Wu, J.; Lefsky, M.A.; Bi, J.; Mantooth, J.A.; Myneni, R.B.; Knyazikhin, Y. Application of Physically-Based Slope Correction for Maximum Forest Canopy Height Estimation Using Waveform Lidar across Different Footprint Sizes and Locations: Tests on LVIS and GLAS. *Remote Sens.* **2014**, *6*, 6566–6586.

# 12

# A PDMS-Based 2-Axis Waterproof Scanner for Photoacoustic Microscopy

**Jin Young Kim [1,†], Changho Lee [2,†], Kyungjin Park [3], Geunbae Lim [1,\*] and Chulhong Kim [1,2,\*]**

[1] Department of Mechanical Engineering, Pohang University of Science and Technology (POSTECH), Pohang 790-784, Korea; E-Mail: ronsan@postech.ac.kr

[2] Department of Creative IT Engineering, Pohang University of Science and Technology (POSTECH), Pohang 790-784, Korea; E-Mail: ch31037@postech.ac.kr

[3] School of Interdisciplinary Bioscience and Bioengineering, Pohang University of Science and Technology (POSTECH), Pohang 790-784, Korea; E-Mail: kjpark@postech.ac.kr

[†] These authors contributed equally to this work.

[\*] Authors to whom correspondence should be addressed; E-Mails: limmems@postech.ac.kr (G.L.); chulhong@postech.ac.kr (C.K.).

Academic Editor: Stefano Mariani

**Abstract:** Optical-resolution photoacoustic microscopy (OR-PAM) is an imaging tool to provide *in vivo* optically sensitive images in biomedical research. To achieve a small size, fast imaging speed, wide scan range, and high signal-to-noise ratios (SNRs) in a water environment, we introduce a polydimethylsiloxane (PDMS)-based 2-axis scanner for a flexible and waterproof structure. The design, theoretical background, fabrication process and performance of the scanner are explained in details. The designed and fabricated scanner has dimensions of $15 \times 15 \times 15$ mm along the X, Y and Z axes, respectively. The characteristics of the scanner are tested under DC and AC conditions. By pairing with electromagnetic forces, the maximum scanning angles in air and water are 18° and 13° along the X and Y axes, respectively. The measured resonance frequencies in air and water are 60 and 45 Hz along the X axis and 45 and 30 Hz along the Y axis, respectively. Finally, OR-PAM with high SNRs is demonstrated using the fabricated scanner, and the PA images of micro-patterned samples and microvasculatures of a mouse ear are successfully obtained with high-resolution and wide-field of view. OR-PAM equipped with the 2-axis PDMS based waterproof scanner has lateral and axial resolutions of 3.6 μm and 26 μm, respectively. This compact OR-PAM system could potentially and widely be used in preclinical and clinical applications.

**Keywords:** optical-resolution photoacoustic microscopy; MEMS scanner; polydimethylsiloxane

---

## 1. Introduction

Photoacoustic microscopy (PAM) has become an important biomedical imaging technique that achieves high-resolution and rich optical contrast by merging optical irradiation and ultrasound detection [1]. When pulsed light illuminates and spreads in biological tissues, targeted biomolecules absorb light and consequently photoacoustic (PA) waves are created via thermoelastic expansion. The produced PA waves are captured by acoustic transducers, and then 1D-, 2D- and/or 3D PA images can be formed to show structural, functional, and molecular information of the tissues [2–6]. Particularly, optical-resolution PAM (OR-PAM) is able to visualize label-free micro-scale images of oxy- and deoxyhemoglobin [7], melanin [8], DNA/RNA in cell nuclei [9], and so on. Due to its strong optical contrast and high optical resolution, OR-PAM has successfully been utilized in many biological studies and applications such as vascular biology [10], neurology [11], ophthalmology [12], intraoperative surgery [13] and so forth. The conventional OR-PAM systems utilize a special opto-ultrasound beam combiner for coaxial and confocal alignment of light illumination and ultrasound detection [14]. This opto-ultrasound beam combiner is the essential part of the conventional OR-PAM systems to reach the maximum signal-to-noise ratios (SNRs). However, the conventional OR-PAM systems suffer from slow imaging speed and are relatively bulky because of linear raster scanning using stepping motors (normally 1 Hz to acquire one depth-resolvable PA B-scan images [15]). However, to expand the usabilities of OR-PAM in preclinical and clinical applications, small size and fast imaging speed are crucial. To improve the imaging speed, microelectromechanical systems (MEMS)-based scanners have been widely developed. Among them silicon-based MEMS scanners have many advantages due to their small sizes and fast scanning speeds. Thus, they have been used in numerous medical imaging applications such as confocal microscopy [16], optical coherence tomography [17], and multi-photon microscopy [18]. Further, the applications have been even more expanded to build handheld probes and endoscopic systems. However, it is difficult to implement the OR-PAM systems with the conventional MEMS scanners because the scanners should be workable in a water environment. In addition, simultaneous co-axial scanning of both light and ultrasound together is another key requirement in order to maintain high SNRs. To address these issues, several approaches have been studied. A 1-axis water immersible scanning mirror has been developed and applied to OR-PAM [19], but it requires an additional bulky linear stage for 2D scanning. Although a 2-axis water-immersible scanning mirror has been introduced [20], the fabrication process includes complex laser cutting of polymer film and manually aligning between two scanning axes. Furthermore, no PA imaging capability has been proved in the report. Recently, our group reports the fast OR-PAM system with a 2-axis water-proofing MEMS scanner [21]. The flowing carbon particles in tube and microvasculatures of a mouse ear are successfully monitored.

In this paper, we introduce design, theoretical background, fabrication and evaluation processes of a 2-axis polydimethylsiloxane (PDMS)-based waterproof scanner (2A-PDMS-WP-scanner) in the previously reported OR-PAM system in details. The soft and water-resistant properties of PDMS make the scanner useable underwater. Moreover, the fabrication process uses micro-milling and soft lithography, which are relatively easy and inexpensive compared to conventional MEMS processes. We test the scanning characteristics of 2A-PDMS-WP-scanner and adopt it in two types of OR-PAM for evaluation of the sensitivity of system. Finally, we have successfully characterized the 2A-PDMS-WP-scanner by photoacoustically imaging gold micro-patterns *in vitro* and microvasculatures of a mouse ear *in vivo*.

## 2. Fabrication and Characterization of a 2A-PDMS-WP Scanner

### 2.1. Design of a 2A-PDMS-WP-Scanner

The 2A-PDMS-WP-scanner mainly consists of a movable structure layer of PDMS and a fixed block of four electromagnets (Figure 1a). The movable structure layer has two scanning axes on a single layer along the X and Y axes (Figure 1b). An aluminum (Al)-coated mirror on the center of the structure layer sufficiently reflects both laser and ultrasound with a reflection rate of 92% and 84%, respectively, in water, and can torsionally be moved along the X and Y axes. As we mentioned before, the key requirement for designing the scanner which can be controllable in water is to overcome the strong damping of torsional oscillation by water and prevent electrical leakage from water. Because of its strong mechanical stability and chemical inertness, PDMS is selected as a material for the structure layer. In addition, PDMS itself is well suited for mechanical sensors or actuators such as accelerometers [22].

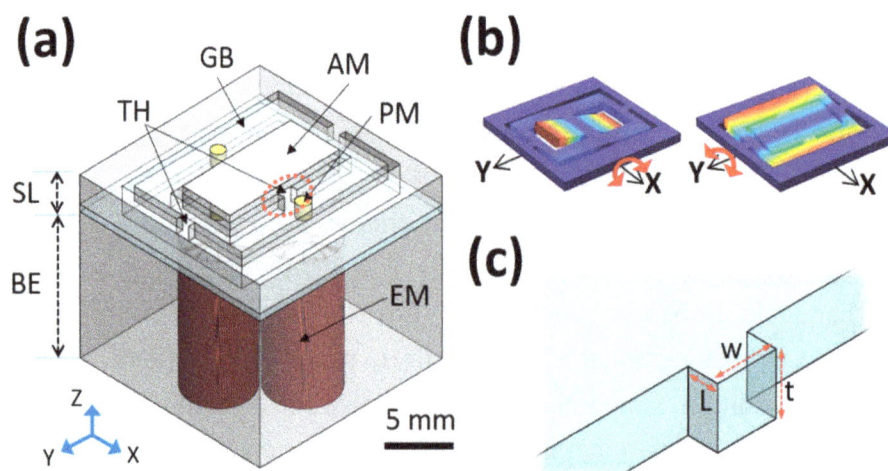

**Figure 1.** Design of the 2A-PDMS-WP-scanner. (**a**) 3D scheme of the 2A-PDMS-WP-scanner; (**b**) Scheme of torsional motions along the X and Y axes; (**c**) Magnified view of a torsional hinge. SL, structure layer; BE, block of four electromagnets; AM, aluminum mirror; PM, permanent magnets; EM, electromagnets; GB, gimbal; and TH, torsional hinges.

The resonant frequency $f_t$, closely related to the scanning speed along the X axis, is calculated as follows:

$$f_t = \frac{1}{2\pi}\sqrt{\frac{K_t}{J}} \tag{1}$$

where $J$ (kg·m$^2$) is the torsional moment of inertia of the mirror, which is calculated as $1 \times 10^{-9}$ kg·m$^2$ by mass and size, and $K_t$ (N·m) is the torsional stiffness of the hinge:

$$K_t = \frac{G}{L}\left[ tw^3 \left\{ \frac{1}{3} - 0.21\frac{w}{t}\left(1 - \frac{w^4}{4t^4}\right)\right\}\right] \tag{2}$$

where $G$ (Pa) is the shear modulus, $L$ (m) is the length, $t$ (m) is the thickness, and $w$ (m) is the width of the torsional hinge (Figure 1c). Because PDMS has a very low $G$ value of ~250 kPa [23], the designed torsional hinge could have a mechanically strong structure with a length $L$ of 0.5 mm, a thickness $t$ of 1 mm, and a width $w$ of 1 mm. Consequently, the calculated torsional stiffness and resonant frequency are 141 µN·m and 59.2 Hz, respectively. If the supporting material has a high $G$ value like the single crystal silicon used in conventional MEMS fabrication (*i.e.*, ~80 GPa), the torsional hinge needs to be very thin to provide a comparable resonant frequency (*i.e.*, a cross section of 40 µm × 40 µm in width and thickness, respectively). Then, the structure becomes delicate and susceptible to damping, and thus it may be fragile when the scanner vibrates in water. In our design, a rigid gimbal acrylic frame and a mirror plate are assembled to the PDMS body to reinforce the elastic PDMS structure except the torsional hinge. The gimbal structure of the single movable layer reduces mechanical coupling between two torsional movements, and thus helps to enhance the scanning preciseness and linearity [24].

Two torsional motions are induced by two independent pairs of electromagnetic forces between four permanent magnets and beneath four electromagnets. The induced magnetic field distribution is simulated using the finite element method magnetics (FEMM) as shown in Figure 2 [25]. When electrical current flows in two oppositely winded enameled coils, the induced magnetic field enables two steel cores be electromagnets. These electromagnet pairs generate attractive and repulsive forces with permanent magnets alternately. Then, the consequently stimulated torque at the torsional hinge induces scanning motion along both X and Y axes.

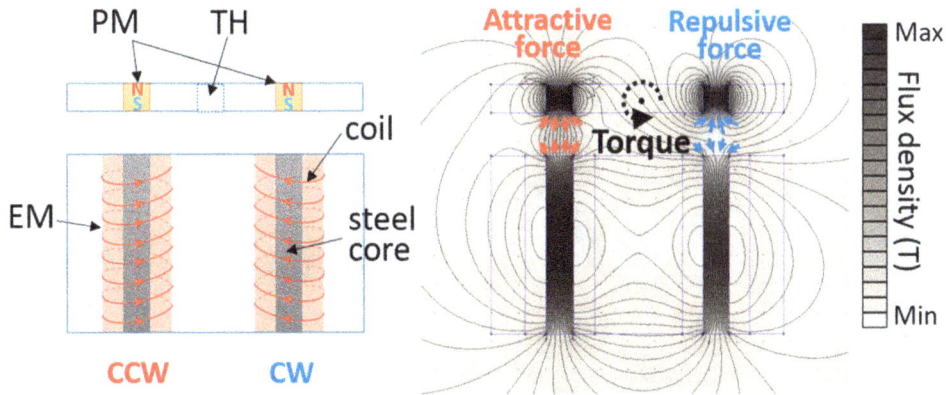

**Figure 2.** Simulated magnetic field distribution induced by electromagnets within the 2A-PDMS-WP-scanner. PM, permanent magnets; EM, electromagnets; TH, torsional hinges; CCW, counter clock wise; and CW, clock wise.

## 2.2. Fabrication Procedures of the 2A-PDMS-WP-Scanner

The movable structure layer of the 2A-PDMS-WP-scanner is fabricated using a PDMS molding process (Figure 3a). PDMS-based soft lithography is another well-developed MEMS process which is widely used in fabrication of microstructures [26]. Using a micro-milling machine, a mold is constructed from an acrylic plate. The end mill with a size of 200 µm is used to cut the smallest body layer with a dimension of 500 µm. The thin acrylic gimbal frames to support and reinforce the flexible PDMS body are also prepared using this milling process. Then, a mixture of PDMS prepolymer and curing agent (Sylgard® 184, Dow Corning, Midland, MI, USA) is prepared with a weight ratio of 10:1, respectively. The mixture is filled into the acrylic mold and cured in a convection oven at 70 °C for 3 h; the resulting solid PDMS body is carefully peeled off from the mold. E-beam evaporation and dicing are used to fabricate the mirror plate by depositing 200-nm-thick Al on the glass plate.

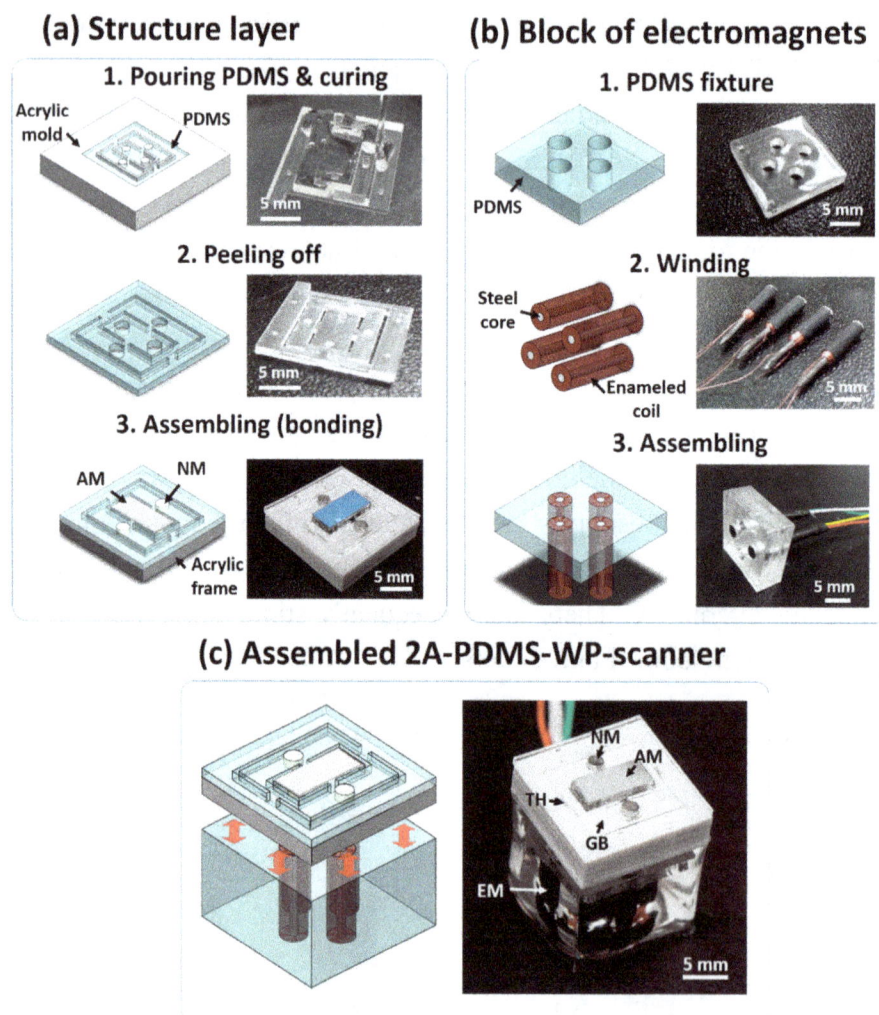

**Figure 3.** Fabrication process of the 2A-PDMS-WP-scanner. (**a**) Structure layer; (**b**) Block of electromagnet; (**c**) The 2A-PDMS-WP-scanner. NM, neodymium magnets; AM, aluminum mirror; TH, torsional hinges; GB, gimbal; and EM, electromagnets.

Finally, the prepared PDMS body, acrylic gimbal frames, four neodymium magnets, and the mirror plate are assembled together using plasma bonding between PDMS and the supporting structures. The same PDMS process is applied to prepare the PDMS fixture to position electromagnets (Figure 3b). A homemade winding system is used to fabricate four electromagnets by winding enameled copper wires with a thickness of 50 μm. These electromagnets are assembled with the PDMS fixture. An additional PDMS molding process is used to engineer a waterproof capsule for the block of electromagnets. Finally, the fabricated structure layer and block of electromagnets are assembled to the 2A-PDMS-WP-scanner using plasma bonding (Figure 3c). The size of the fabricated 2A-PDMS-WP-scanner is 15 × 15 × 15 mm along X, Y and Z axes, respectively.

## 2.3. Scanning Characteristics of the 2A-PDMS-WP-Scanner

Two important scanning characteristics of the 2A-PDMS-WP-scanner include the maximum scanning angle and scanning speed; these characteristics are related to the field of view (FOV) and imaging speed in OR-PAM, respectively. These characteristics are determined by the applied amplitude and frequency of driving signals. The driving signals are generated by a data acquisition (DAQ) system (NI PCIe-6321, National Instruments, Austin, TX, USA) and amplified by a high-current operational amplifier (OPA2544, Texas Instruments, Dallas, TX, USA). The scanning angle can be calculated by measuring the scanning length from the center of the mirror plate to the preset positions.

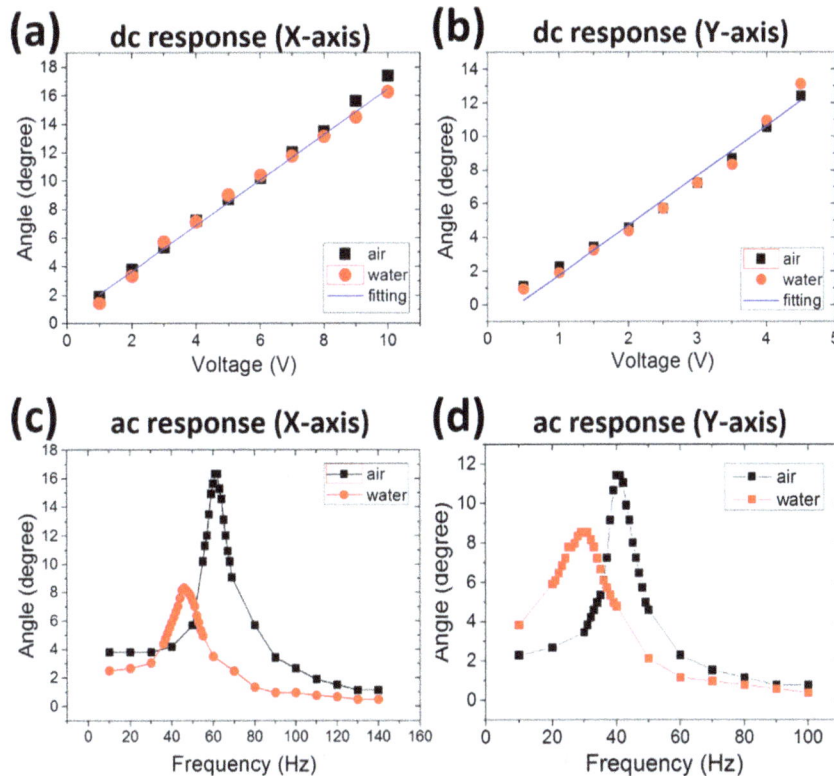

**Figure 4.** Scanning responses of the 2A-WP-MEMS-scanner in DC and ac conditions. Scanning angles *versus* the applied DC voltages along the (**a**) X and Y (**b**) axes, respectively. Scanning angles *versus* the applied ac frequencies at 2 V along (**c**) X and (**d**) Y axes, respectively.

When DC voltages are applied, the scanning angles increase linearly with the applied voltages in both X and Y axes as shown in Figure 4a,b. Further, the scanning angles in both axes are almost identical in both water and air environments. The maximum scanning angles are 18° at 10 V (Figure 4a) and 13° at 5 V along the X and Y axes (Figure 4b). The corresponding FOVs are 14.2 mm × 10.2 mm along the X and Y axes, respectively, at 22 mm away from the mirror surface where the typical light and ultrasound foci are. Our results imply that the fabricated 2A-PDMS-WP-scanner provides the wide FOV even with DC voltage application due to its low torsional stiffness of the hinge. Although the scanner can work well under the DC condition, the application of an AC voltage gives a fast and wide scanning capability when the scanning speed is near the resonant frequency (Figure 4c,d). In air, the resonance frequencies are 60 and 45 Hz along the X and Y axes, respectively. In water, the resonance frequencies are 45 and 30 Hz along the X and Y axes, respectively, due to damping by water. The damping effect reduces the scanning angle (*i.e.*, 8.5° and 8.5° at 2V along X and Y axes, respectively) in water compared to that in air. The corresponding FOVs are 6.6 mm × 6.6 mm along the X and Y axes, respectively, at 22 mm away from the mirror surface where our targets are.

### 3. *In Vitro* and *in Vivo* Photoacoustic Imaging Using the 2A-WP-MEMS-Scanner

#### 3.1. Comparison of Signal-to-Noise Ratios (SNRs)

Two forms of the OR-PAM systems are implemented using the 2A-PDMS-WP-scanner (Figure 5). A Q-switched-diode-pumped-solid-state laser (SPOT-10-200-532, Elforlight, Daventry, UK) is commonly used to deliver a 532-nm laser beam (repetition rate, 10 kHz) in all experiments. The first form (OR-PAM I) is developed by only scanning the focused laser beam via an objective lens (AC254-060-A, Thorlabs, Newton, NJ, USA) and the 2A-PDMS-WP-scanner (Figure 5a) [27]. In this mode, an unfocused transducer (10 MHz center frequency, V312, Olympus NDT, Waltham, MA, USA) is utilized to detect the PA waves. The second form (OR-PAM II) requires an additional opto-ultrasound beam combiner to achieve simultaneous confocal scanning of laser and ultrasound (Figure 5b). The focused laser beam is first reflected by an Al film between two right angled prisms in the beam combiner, and then directed to the surface of a sample via the 2A-PDMS-WP-scanner. The induced PA waves are first reflected by the scanner, passed through the opto-ultrasound beam combiner, and detected by an ultrasound transducer (50 MHz center frequency, V214-BB-RM, Olympus NDT). Interestingly, the Al coating of the scanner surface enables to reflect both light and ultrasound, while the Al film between two prisms enables to reflect only light and is transparent to ultrasound. Further, the attached acoustic lens (NT45-384, Edmund, Barrington, NJ, USA) in front of the opto-ultrasound beam combiner can manipulate focused ultrasound. The imaging speeds of both types are identical because they ultimately rely on the laser repetition rate and the scanning speed. However, the SNRs of the second form is much better than that of the first one because the coaxial geometry of light and ultrasound boost the SNRs. By imaging a piece of black vinyl tape, we are able to compare the SNRs of both systems. The Hilbert-transformed PA A-line profiles acquired by the OR-PAM I and II systems are shown in Figure 5c. The quantified SNRs measured by the OR-PAM I and II systems are 25 and 39 dB, respectively. Our results imply that the coaxial scanning approach of light and ultrasound using the OR-PAM II equipped with 2A-PDMS-WP-scanner (*i.e.*, called as 2A-PDMS-WP-OR-PAM) sufficiently improve the SNRs while maintaining the imaging speed, which is a key requirement for *in vivo* real-time PA imaging.

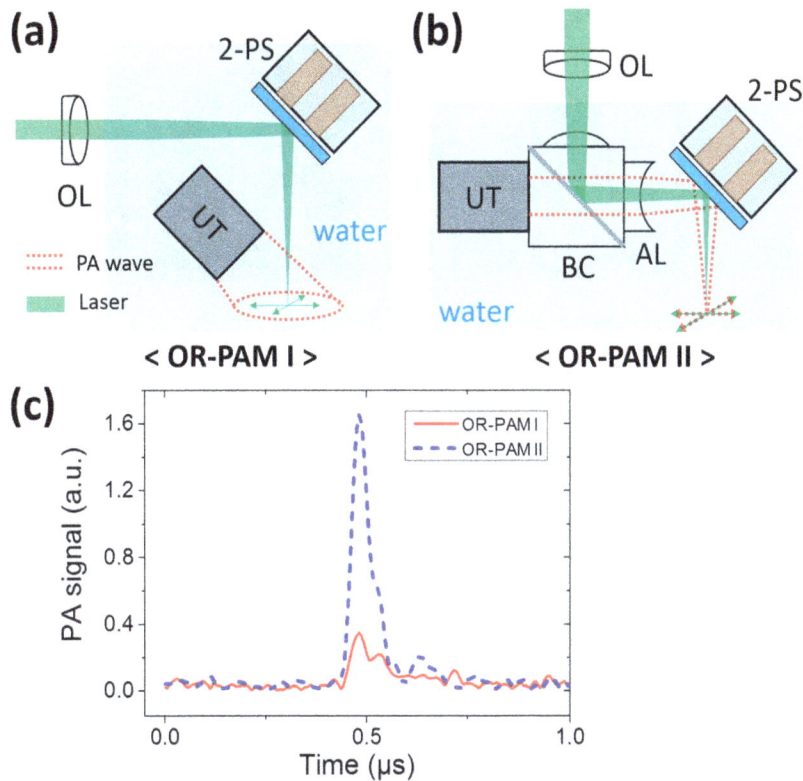

**Figure 5.** Comparison of signal-to-noise ratios in two forms of OR-PAM. (**a**) Schematic of the OR-PAM I system. An unfocused ultrasound transducer is used and the laser beam is only scanned by the 2A-PDMS-WP-scanner; (**b**) Schematic of the OR-PAM II system. Both light and ultrasound are confocally scanned by the 2A-PDMS-WP-scanner; (**c**) Hilbert-transformed PA A-line profiles of a black vinyl tape acquired by the OR-PAM I and OR-PAM II systems, respectively. UT, ultrasonic transducer; OL, objective lens; BC, beam combiner; AL, acoustic lens; and 2-PS, 2A-PDMS-WP-scanner.

## 3.2. In Vitro Photoacoustic Imaging of Gold Micro-Patterns

To verify the performance of the 2A-PDMS-WP-OR-PAM system, a gold micro-patterned sample with a thickness of 200 nm on a glass plate (Figure 6a) is photoacoustically imaged (Figure 6b) *in vitro*. The gold micro-patterns are prepared by a conventional MEMS process, which includes a photolithography for micro-patterning and etching the gold film on a glass substrate. The smallest width of the gold micro-patterns is 20 μm on the center. The scanning range is 2 mm × 1.8 mm along the X and Y axes, respectively. As shown in Figure 6b, the gold micro-patterns are clearly visualized in the PA image. To quantify the lateral resolution, the PA amplitude profile along the a-a' line is plotted in Figure 6c. The line spread function (LSF) is derived from the fitting curve of the edge spread function (ESF) in the range of from 0 to 18 μm. The calculated full width at half maximum (FWHM) of the LSF, a lateral resolution of the system, is 3.6 μm, and the value well matches with the theoretical estimation (*i.e.*, 3.4 μm). The axial resolution is determined by the one-way acoustic bandwidth, approximately 26 μm [15].

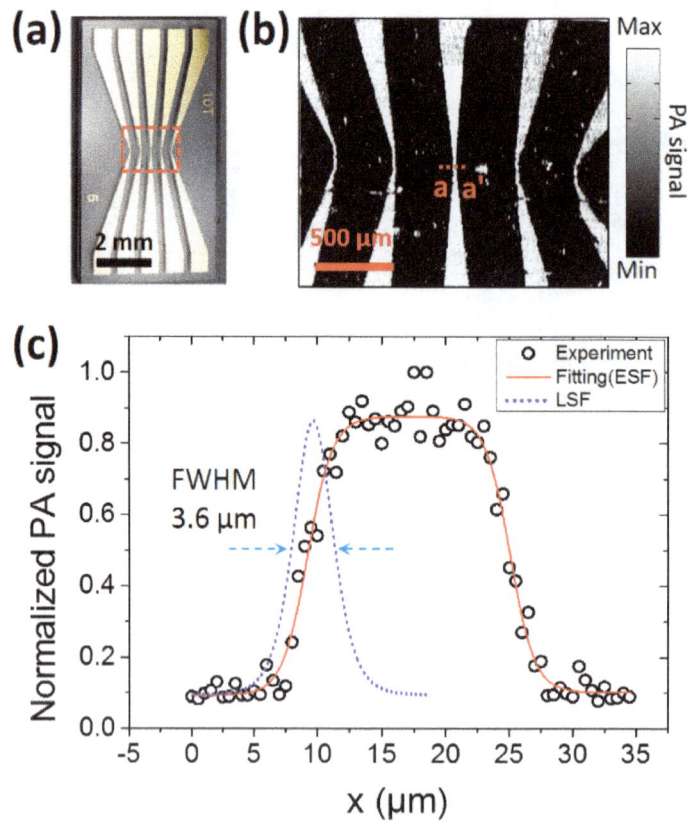

**Figure 6.** *In vitro* PA imaging of gold micro-patterns using the 2A-PDMS-WP-OR-PAM. (**a**) Photograph of gold micro-patterns; (**b**) PA maximum amplitude projection image of the red boxed region of (a); (**c**) PA amplitude profile across the a-a′ line of (b) and fitted edge spread function (ESF) and line spread function (LSF). Full width at half maximum (FWHM) in the plot means a lateral resolution of the system.

### 3.3. In Vivo Photoacoustic Imaging of a Mouse Ear

Lastly, *in vivo* PA imaging of a mouse ear is conducted to verify the high SNRs, wide FOV, and fast scanning of the 2A-PDMS-WP-OR-PAM system. All animal experimental procedures satisfy with the laboratory animal protocol admitted by the institutional animal care and use committee of Pohang University of Science and Technology (POSTECH). A healthy Balb/c mouse (weighing ~20 g) is initially anesthetized and maintained with vaporized isoflurane gas. The fine hair on the mouse ear is removed by commercial depilatory, and then the mouse is positioned on a handmade animal holder. An electrical heating pad is used to keep the body temperature. The excited pulsed laser energy (*i.e.*, 13 mJ/cm$^2$) is below the safety limit (*i.e.*, ~20 mJ/cm$^2$) of laser usage on skins. To acquire dense pixels numbers (*i.e.*, 1000 pixels × 500 pixels along the X and Y axes, respectively), a triangular waveform with a frequency of 5 Hz and a voltage of 5 V is used for the B-scan image along the X axis. A sawtooth waveform with a frequency of 0.01 Hz and a voltage of 5 V is used for volumetric imaging along the Y axis. A volumetric PA image with the wide FOV (*i.e.*, 6.6 mm × 11 mm along the X and Y axes, respectively) is displayed within 100 seconds. Figure 7a shows the photograph of the mouse ear and the corresponding *in vivo* PA image is shown in Figure 7b. The PA image clearly visualizes large blood vessels as well as capillaries. The imaging speed of the 2A-PDMS-WP-OR-PAM system has been currently limited by the repetition rate of the used pulsed laser. We expect that volumetric PA imaging rate will be further enhanced by using faster pulse laser (*i.e.*, 500 kHz) and the optimized imaging processing.

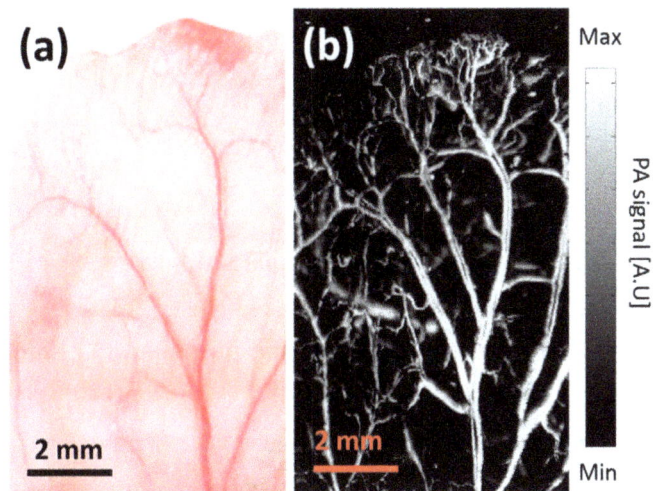

**Figure 7.** *In vivo* noninvasive PA imaging of the microvasculatures in a mouse ear. (**a**) Photograph of the mouse ear; (**b**) *In vivo* corresponding noninvasive PA image of the mouse ear.

## 4. Conclusions

We have developed a 2-axis PDMS-based waterproof scanner and demonstrate its use in two forms of OR-PAM systems. The fabrication process of the scanner is detailed in each step. The main approach is soft lithography of PDMS, which has low stiffness and good water resistance. The fabricated scanner simultaneously and coaxially reflect both ultrasound and light in both air and water. The small size and simple operation using DC or AC driving voltages make the PDMS scanner versatile in different forms of OR-PAM systems. A gold micro-pattern is photoacousticlly imaged *in vitro* and finally, we successfully image microvasculatures of a mouse ear *in vivo* using the developed OR-PAM systems. This 2A-PDMS-WP-scanner can be potentially used for many types of OR-PAM systems such as handheld probes, endoscopic probes, laparoscopic probes, and so on.

## Acknowledgments

This work was supported by the research funds from the IITP IT Consilience Creative Program (IITP-2015-R0346-15-1007), the NRF Engineering Research Center grant (NRF-2011-0030075), and the NRF Mid-career Researcher Program (NRF-2012R1A2A2A06047424) of the Ministry of Science, ICT and Future Planning, Republic of Korea.

## Author Contributions

J.Y.K. designed and fabricated the PDMS scanner and C.L. designed and developed the two types of PAM systems; they also designed and performed *in vitro* and *in vivo* PA imaging, collected samples, analyzed and interpreted data, prepared the figures for the manuscript, and wrote the manuscript. K.P. contributed to characterization of the scanner and electromagnetic simulation test.

G.L. and C.K. conceived and supervised the project, designed experiments, interpreted data, and wrote the manuscript. All authors contributed to critical reading of the manuscript.

**Conflicts of Interest**

The authors declare no conflict of interest.

**References**

1. Kim, C.; Favazza, C.; Wang, L.V. *In vivo* photoacoustic tomography of chemicals: High-resolution functional and molecular optical imaging at new depths. *Chem. Rev.* **2010**, *110*, 2756–2782.
2. Jeon, M.; Kim, J.; Kim, C. Multiplane spectroscopic whole-body photoacoustic imaging of small animals *in vivo*. *Med. Biol. Eng. Comput.* **2014**, 1–12.
3. Zhang, Y.; Jeon, M.; Rich, L.J.; Hong, H.; Geng, J.; Zhang, Y.; Shi, S.; Barnhart, T.E.; Alexandridis, P.; Huizinga, J.D.; *et al.* Non-invasive multimodal functional imaging of the intestine with frozen micellar naphthalocyanines. *Nat. Nanotechnol.* **2014**, *9*, 631–638.
4. Liu, X.; Law, W.-C.; Jeon, M.; Wang, X.; Liu, M.; Kim, C.; Prasad, P.N.; Swihart, M.T. $Cu_{2-x}Se$ nanocrystals with localized surface plasmon resonance as sensitive contrast agents for *in vivo* photoacoustic imaging: Demonstration of sentinel lymph node mapping. *Adv. Healthc. Mater.* **2013**, *2*, 952–957.
5. Srivatsan, A.; Jenkins, S.V.; Jeon, M.; Wu, Z.; Kim, C.; Chen, J.; Pandey, R.K. Gold nanocage-photosensitizer conjugates for dual-modal image-guided enhanced photodynamic therapy. *Theranostics* **2014**, *4*, 163–174.
6. Kim, C.; Jeon, M.; Wang, L.V. Nonionizing photoacoustic cystography *in vivo*. *Opt. Lett.* **2011**, *36*, 3599–3601.
7. Lee, C.; Jeon, M.; Jeon, M.Y.; Kim, J.; Kim, C. *In vitro* photoacoustic measurement of hemoglobin oxygen saturation using a single pulsed broadband supercontinuum laser source. *Appl. Opt.* **2014**, *53*, 3884–3889.
8. Wang, Y.; Maslov, K.; Zhang, Y.; Hu, S.; Yang, L.; Xia, Y.; Liu, J.; Wang, L.V. Fiber-laser-based photoacoustic microscopy and melanoma cell detection. *J. Biomed. Opt.* **2011**, *16*, doi:10.1117/1.3525643.
9. Yao, D.-K.; Maslov, K.; Shung, K.K.; Zhou, Q.; Wang, L.V. *In vivo* label-free photoacoustic microscopy of cell nuclei by excitation of DNA and RNA. *Opt. Lett.* **2010**, *35*, 4139–4141.
10. Cai, X.; Zhang, Y.; Li, L.; Choi, S.-W.; MacEwan, M.R.; Yao, J.; Kim, C.; Xia, Y.; Wang, L.V. Investigation of neovascularization in three-dimensional porous scaffolds *in vivo* by a combination of multiscale photoacoustic microscopy and optical coherence tomography. *Tissue Eng. Part C Methods* **2012**, *19*, 196–204.
11. Hu, S.; Wang, L.V. Neurovascular photoacoustic tomography. *Front. Neuroeng.* **2010**, *2*, doi:10.3389/fnene.2010.00010.
12. Jiao, S.; Jiang, M.; Hu, J.; Fawzi, A.; Zhou, Q.; Shung, K.K.; Puliafito, C.A.; Zhang, H.F. Photoacoustic ophthalmoscopy for *in vivo* retinal imaging. *Opt. Express* **2010**, *18*, 3967–3972.
13. Han, S.; Lee, C.; Kim, S.; Jeon, M.; Kim, J.; Kim, C. *In vivo* virtual intraoperative surgical photoacoustic microscopy. *Appl. Phys. Lett.* **2013**, *103*, doi:10.1063/1.4830045.

14. Hu, S.; Maslov, K.; Wang, L.V. Second-generation optical-resolution photoacoustic microscopy with improved sensitivity and speed. *Opt. Lett.* **2011**, *36*, 1134–1136.

15. Yao, J.; Wang, L.V. Photoacoustic microscopy. *Laser Photonics Rev.* **2013**, *7*, 758–778.

16. Piyawattanametha, W.; Hyejun, R.; Mandella, M.J.; Loewke, K.; Wang, T.D.; Kino, G.S.; Solgaard, O.; Contag, C.H. 3-D near-infrared fluorescence imaging using an mems-based miniature dual-axis confocal microscope. *J. Sel. Top. Quantum Electron.* **2009**, *15*, 1344–1350.

17. Pan, Y.; Xie, H.; Fedder, G.K. Endoscopic optical coherence tomography based on a microelectromechanical mirror. *Opt. Lett.* **2001**, *26*, 1966–1968.

18. Jung, W.; Tang, S.; McCormic, D.T.; Xie, T.; Ahn, Y.-C.; Su, J.; Tomov, I.V.; Krasieva, T.B.; Tromberg, B.J.; Chen, Z. Miniaturized probe based on a microelectromechanical system mirror for multiphoton microscopy. *Opt. Lett.* **2008**, *33*, 1324–1326.

19. Yao, J.; Huang, C.-H.; Wang, L.; Yang, J.-M.; Gao, L.; Maslov, K.I.; Zou, J.; Wang, L.V. Wide-field fast-scanning photoacoustic microscopy based on a water-immersible mems scanning mirror. *J. Biomed. Opt.* **2012**, *17*, 0805051–0805053.

20. Huang, C.-H.; Yao, J.; Wang, L.; Zou, J. A water-immersible 2-axis scanning mirror microsystem for ultrasound andha photoacoustic microscopic imaging applications. *Microsyst. Technol.* **2013**, *19*, 577–582.

21. Kim, J.Y.; Lee, C.; Park, K.; Lim, G.; Kim, C. Fast optical-resolution photoacoustic microscopy using a 2-axis water-proofing mems scanner. *Sci. Rep.* **2015**, *5*, doi:10.1038/srep07932.

22. Joost, C.L.; Wouter, O.; Peter, H.V.; Piet, B. Polydimethylsiloxane as an elastic material applied in a capacitive accelerometer. *J. Micromech. Microeng.* **1996**, *6*, doi:10.1088/0960-1317/6/1/010.

23. Lötters, J.C.; Olthuis, W.; Veltink, P.H.; Bergveld, P. The mechanical properties of the rubber elastic polymer polydimethylsiloxane for sensor applications. *J. Micromech. Microeng.* **1997**, *7*, doi:10.1088/0960-1317/7/3/017.

24. Acar, C.; Shkel, A. *Mems Vibratory Gyroscopes, Structural Approaches to Improve Robustness*; Springer: New York, NY, USA, 2009.

25. Meeker, D.C. Finite Element Method Magnetics. Available online: http://www.femm.info/ (accessed on 22 April 2015).

26. Xia, Y.; Whitesides, G.M. Soft lithography. *Angew. Chem. Int. Ed.* **1998**, *37*, 550–575.

27. Xie, Z.; Jiao, S.; Zhang, H.F.; Puliafito, C.A. Laser-scanning optical-resolution photoacoustic microscopy. *Opt. Lett.* **2009**, *34*, 1771–1773.

# Permissions

The contributors of this book come from diverse backgrounds, making this book a truly international effort. This book will bring forth new frontiers with its revolutionizing research information and detailed analysis of the nascent developments around the world.

We would like to thank all the contributing authors for lending their expertise to make the book truly unique. They have played a crucial role in the development of this book. Without their invaluable contributions this book wouldn't have been possible. They have made vital efforts to compile up to date information on the varied aspects of this subject to make this book a valuable addition to the collection of many professionals and students.

This book was conceptualized with the vision of imparting up-to-date information and advanced data in this field. To ensure the same, a matchless editorial board was set up. Every individual on the board went through rigorous rounds of assessment to prove their worth. After which they invested a large part of their time researching and compiling the most relevant data for our readers.

The editorial board has been involved in producing this book since its inception. They have spent rigorous hours researching and exploring the diverse topics which have resulted in the successful publishing of this book. They have passed on their knowledge of decades through this book. To expedite this challenging task, the publisher supported the team at every step. A small team of assistant editors was also appointed to further simplify the editing procedure and attain best results for the readers.

Apart from the editorial board, the designing team has also invested a significant amount of their time in understanding the subject and creating the most relevant covers. They scrutinized every image to scout for the most suitable representation of the subject and create an appropriate cover for the book.

The publishing team has been an ardent support to the editorial, designing and production team. Their endless efforts to recruit the best for this project, has resulted in the accomplishment of this book. They are a veteran in the field of academics and their pool of knowledge is as vast as their experience in printing. Their expertise and guidance has proved useful at every step. Their uncompromising quality standards have made this book an exceptional effort. Their encouragement from time to time has been an inspiration for everyone.

The publisher and the editorial board hope that this book will prove to be a valuable piece of knowledge for researchers, students, practitioners and scholars across the globe.

# List of Contributors

**Ambra Giannetti**
CNR—Institute of Applied Physics "Nello Carrara", Via Madonna del Piano 10, 50019 Sesto Fiorentino (FI), Italy

**Andrea Barucci**
CNR—Institute of Applied Physics "Nello Carrara", Via Madonna del Piano 10, 50019 Sesto Fiorentino (FI), Italy

**Franco Cosi**
CNR—Institute of Applied Physics "Nello Carrara", Via Madonna del Piano 10, 50019 Sesto Fiorentino (FI), Italy

**Stefano Pelli**
CNR—Institute of Applied Physics "Nello Carrara", Via Madonna del Piano 10, 50019 Sesto Fiorentino (FI), Italy
Museo Storico della Fisica e Centro Studi e Ricerche Enrico Fermi, Piazza del Viminale 1, 00184 Rome, Italy

**Sara Tombelli**
CNR—Institute of Applied Physics "Nello Carrara", Via Madonna del Piano 10, 50019 Sesto Fiorentino (FI), Italy

**Cosimo Trono**
CNR—Institute of Applied Physics "Nello Carrara", Via Madonna del Piano 10, 50019 Sesto Fiorentino (FI), Italy

**Francesco Baldini**
CNR—Institute of Applied Physics "Nello Carrara", Via Madonna del Piano 10, 50019 Sesto Fiorentino (FI), Italy

**Antonio Bueno**
Service d'Electromagnétisme et de Télécommunications, Université de Mons, Boulevard Dolez 31, 7000 Mons, Belgium

**Driss Lahem**
Materia Nova, Materials R&D Centre, Parc Initialis, Avenue Nicolas Copernic 1, 7000 Mons, Belgium

**Christophe Caucheteur**
Service d'Electromagnétisme et de Télécommunications, Université de Mons, Boulevard Dolez 31, 7000 Mons, Belgium

**Marc Debliquy**
Service de Science des Matériaux, Université de Mons, Rue de l'Epargne 56, 7000 Mons, Belgium

**Simone Brienza**
Department of Information Engineering, University of Pisa, Largo Lucio Lazzarino 1, 56122 Pisa, Italy

**Andrea Galli**
Department of Information Engineering, University of Pisa, Largo Lucio Lazzarino 1, 56122 Pisa, Italy

**Giuseppe Anastasi**
Department of Information Engineering, University of Pisa, Largo Lucio Lazzarino 1, 56122 Pisa, Italy

**Paolo Bruschi**
Department of Information Engineering, University of Pisa, Via G. Caruso 16, 56122 Pisa, Italy

**Yu Rang Park**
Clinical Research Center, Asan Medical Center, Seoul 138-736, Korea

**Yura Lee**
Department of Biomedical Informatics, Asan Medical Center, Seoul 138-736, Korea

**Guna Lee**
Division of Nursing Science, College of Health Science, Ewha Womans University, Seoul 120-750

**Jae Ho Lee**
Department of Biomedical Informatics, Asan Medical Center, Seoul 138-736, Korea
Ubiquitous Health Center, Asan Medical Center, Seoul 138-736, Korea
Department of Emergency Medicine, Asan Medical Center, University of Ulsan College of Medicine, Seoul 138-736, Korea
Division of General Internal Medicine and Primary Care, Brigham and Women's Hospital, Boston, MA 02467, USA

**Soo-Yong Shin**
Department of Biomedical Informatics, Asan Medical Center, Seoul 138-736, Korea
Ubiquitous Health Center, Asan Medical Center, Seoul 138-736, Korea

**Katja Herzog**
Julius Kühn-Institut-Federal Research Centre of Cultivated Plants, Institute for Grapevine Breeding Geilweilerhof, Siebeldingen 76833, Germany

**Rolf Wind**
Julius Kühn-Institut-Federal Research Centre of Cultivated Plants, Institute for Grapevine Breeding Geilweilerhof, Siebeldingen 76833, Germany

**Reinhard Töpfer**
Julius Kühn-Institut-Federal Research Centre of Cultivated Plants, Institute for Grapevine Breeding Geilweilerhof, Siebeldingen 76833, Germany

**Kun Xiong**
Key Laboratory of Precision Opto-Mechatronics Technology, Ministry of Education, School of Instrumentation Science and Opto-Electronics Engineering, Beijing University of Aeronautics and Astronautics (BUAA), Beijing 100191, China

**Jie Jiang**
Key Laboratory of Precision Opto-Mechatronics Technology, Ministry of Education, School of Instrumentation Science and Opto-Electronics Engineering, Beijing University of Aeronautics and Astronautics (BUAA), Beijing 100191, China

**Zengfeng Du**
Optics and Optoelectronics Laboratory, Ocean University of China, Qingdao 266100, China

**Jing Chen**
Optics and Optoelectronics Laboratory, Ocean University of China, Qingdao 266100, China

**Wangquan Ye**
Optics and Optoelectronics Laboratory, Ocean University of China, Qingdao 266100, China

**Jinjia Guo**
Optics and Optoelectronics Laboratory, Ocean University of China, Qingdao 266100, China

**Xin Zhang**
Key Lab of Marine Geology and Environment, Institute of Oceanology, Chinese Academy of Sciences, Qingdao 266071, China

**Ronger Zheng**
Optics and Optoelectronics Laboratory, Ocean University of China, Qingdao 266100, China

**Asier Moreno**
Deusto Institute of Technology (DeustoTech), University of Deusto, Bilbao 48007, Spain

**Asier Perallos**
Deusto Institute of Technology (DeustoTech), University of Deusto, Bilbao 48007, Spain

**Diego López-de-Ipiña**
Deusto Institute of Technology (DeustoTech), University of Deusto, Bilbao 48007, Spain

**Enrique Onieva**
Deusto Institute of Technology (DeustoTech), University of Deusto, Bilbao 48007, Spain

**Itziar Salaberria**
Deusto Institute of Technology (DeustoTech), University of Deusto, Bilbao 48007, Spain

**Antonio D. Masegosa**
Deusto Institute of Technology (DeustoTech), University of Deusto, Bilbao 48007, Spain
IKERBASQUE, Basque Foundation for Science, Bilbao 48011, Spain

**Hongyi Ge**
State Key Laboratory of Transducer Technology, Institute of Electronics, Chinese Academy of Sciences, Beijing 100080, China
University of the Chinese Academy of Sciences, Beijing 100080, China

**Yuying Jiang**
State Key Laboratory of Transducer Technology, Institute of Electronics, Chinese Academy of Sciences, Beijing 100080, China
University of the Chinese Academy of Sciences, Beijing 100080, China

**Feiyu Lian**
Key Laboratory of Grain Information Processing & Control, Ministry of Education, Zhengzhou 450001, China

**Yuan Zhang**
Key Laboratory of Grain Information Processing & Control, Ministry of Education, Zhengzhou 450001, China

**Shanhong Xia**
State Key Laboratory of Transducer Technology, Institute of Electronics, Chinese Academy of Sciences, Beijing 100080, China
University of the Chinese Academy of Sciences, Beijing 100080, China

**Hanguen Kim**
Urban Robotics Laboratory (URL), Dept. Civil and Environmental Engineering, Korea Advanced Institute of Science and Technology (KAIST), 291 Daehak-ro, Yuseong-gu, Daejeon 305-338, Korea

**Sangwon Lee**
Urban Robotics Laboratory (URL), Dept. Civil and Environmental Engineering, Korea Advanced Institute of Science and Technology (KAIST), 291 Daehak-ro, Yuseong-gu, Daejeon 305-338, Korea

**Dongsung Lee**
Image & Video Research Group, Samsung S1 Cooperation, 168 S1 Building, Soonhwa-dong, Joong-gu, Seoul 100-773, Korea

**Soonmin Choi**
Image & Video Research Group, Samsung S1 Cooperation, 168 S1 Building, Soonhwa-dong, Joong-gu, Seoul 100-773, Korea

**Jinsun Ju**
Image & Video Research Group, Samsung S1 Cooperation, 168 S1 Building, Soonhwa-dong, Joong-gu, Seoul 100-773, Korea

**Hyun Myung**
Urban Robotics Laboratory (URL), Dept. Civil and Environmental Engineering, Korea Advanced Institute of Science and Technology (KAIST), 291 Daehak-ro, Yuseong-gu, Daejeon 305-338, Korea

**Zhugeng Duan**
Key Laboratory of Digital Earth Science, Institute of Remote Sensing and Digital Earth (RADI), Chinese Academy of Science, Haidian District, Beijing 100094, China
School of GeoSciences and Info-Physics, Central South University, Changsha 410083, China
School of Sciences, Central South University of Forestry and Technology, Changsha 410004, China

**Dan Zhao**
Key Laboratory of Digital Earth Science, Institute of Remote Sensing and Digital Earth (RADI), Chinese Academy of Science, Haidian District, Beijing 100094, China

**Yuan Zeng**
Key Laboratory of Digital Earth Science, Institute of Remote Sensing and Digital Earth (RADI), Chinese Academy of Science, Haidian District, Beijing 100094, China

**Yujin Zhao**
Key Laboratory of Digital Earth Science, Institute of Remote Sensing and Digital Earth (RADI), Chinese Academy of Science, Haidian District, Beijing 100094, China

**Bingfang Wu**
Key Laboratory of Digital Earth Science, Institute of Remote Sensing and Digital Earth (RADI), Chinese Academy of Science, Haidian District, Beijing 100094, China

**Jianjun Zhu**
School of GeoSciences and Info-Physics, Central South University, Changsha 410083, China

**Jin Young Kim**
Department of Mechanical Engineering, Pohang University of Science and Technology (POSTECH), Pohang 790-784, Korea

**Changho Lee**
Department of Creative IT Engineering, Pohang University of Science and Technology (POSTECH), Pohang 790-784, Korea

**Kyungjin Park**
School of Interdisciplinary Bioscience and Bioengineering, Pohang University of Science and Technology (POSTECH), Pohang 790-784, Korea

**Geunbae Lim**
Department of Mechanical Engineering, Pohang University of Science and Technology (POSTECH), Pohang 790-784, Korea

**Chulhong Kim**
Department of Mechanical Engineering, Pohang University of Science and Technology (POSTECH), Pohang 790-784, Korea
Department of Creative IT Engineering, Pohang University of Science and Technology (POSTECH), Pohang 790-784, Korea